ADVANCES IN GAS CHROMATOGRAPHY

Edited by **Xinghua Guo**

Advances in Gas Chromatography

http://dx.doi.org/10.5772/57016

Edited by Xinghua Guo

Contributors

Serban Moldoveanu, Jesuí Visentainer, Swami Maruyama, Thiago Claus, Oscar Santos Junior, Lucas Chiavelli, James Lewicki, Robert Maxwell, Yunping Qiu, Deborah Reed, Maria Teresa Teresa Nunez-Cardona, Ernesto Bernal, Elena Stashenko, Jairo René Martínez, Sokhansanj Shahab, Fahimeh Yazdanpanah

Published by InTech

Janeza Trdine 9, 51000 Rijeka, Croatia

Publishing Process Manager Iva Lipovic

Technical Editor InTech DTP team

Cover InTech Design team

First published February, 2014

Printed in Croatia

Additional hard copies can be obtained from orders@intechopen.com

Advances in Gas Chromatography , Edited by Xinghua Guo

p. cm.

ISBN 978-953-51-1227-3

For EU product safety concerns:
IN TECH d.o.o., Prolaz Marije Krucifikse Kozulić 3, 51000 Rijeka, Croatia, info@intechopen.com or visit our website at intechopen.com.

Meet the editor

Xinghua Guo obtained his Ph.D. in Mass Spectrometry/ Chemistry at the Chinese Academy of Sciences, China, before he moved to Germany (Alexander von-Humboldt Foundation fellow), the Netherlands and Austria. He has been working in the field of instrumental analysis, using a broad range of MS, LC-MS and GC-MS, with interests in various method developments, ranging from instrumental fundamentals, applications and organic sample preparation. Since 2006 he has worked as Assistant Professor at the Graz University of Technology (Austria), where he obtained the habilitation status (Uni. Doz.) in Analytical Chemistry, gave lectures in 'LC-MS Bioanalysis' and 'Organic Instrumental Analysis', and supervised several dissertations. Currently he is specializing in characterization of small molecules in biosimilar development in the Analytical Characterization Group at Sandoz Biopharmaceuticals, Kundl, Austria. Xinghua has (co-)authored over 70 scientific publications and 2 patents.

Contents

Preface

The steady development and improvement of gas chromatographic instrumentations and applications in combination with advanced detection techniques has been leading GC techniques to one of the ideal methods of choices since decades. It will remain an irreplaceable analytical technique in many research areas for both quantitative analysis and qualitative characterization/identification, although it has been facing a dramatic impact as a result of the rapid development of liquid chromatography-mass spectrometry. With this book it is not the intention to cover the field comprehensively, but to highlight a few areas where significant advances have been reported recently and/or a revisit of basic concepts is deserved. It provides an overview of selected instrumentation developments, frontline and modern research as well as recent industrial applications, aiming at providing a practical reference in daily applications.

First a review on recent developments of GC-MS is given, followed by a practical guidance of quantification of fatty acids using GC-FID and determination of detection limits in general. Chapter 4 critically reviews one of the ever expanding research areas: GC-based metabolomics applications in biomedical, plant and microbial research. It covers aspects from sample preparation, derivatization, data processing and interpretation to application areas. Furthermore, Chapter 5 illustrates the power of GC-MS profiling of antioxidants in botanicals with practical examples. On the other hand, Chapter 7 focuses on the applications of pyrolysis GC-MS for the study of degradation and aging of complex synthetic silicone elastomers, which demonstrates the advantages and limitations of the technique in material sciences.

This book is a collection of contributions of several experts in their representing fields, for which I am very grateful to all of them. It contains useful hand-on practices of fundamentals, sample preparation (derivatization) and data processing in daily research. This book is recommended to both basic and experienced researchers in gas chromatography.

Dr. Xinghua Guo
Uni.-Doz.
Analytical Characterization
Biopharmaceuticals
Sandoz GmbH
Kundl, Austria

Chapter 1

Gas Chromatography-Mass Spectrometry

Elena Stashenko and Jairo René Martínez

Additional information is available at the end of the chapter

http://dx.doi.org/10.5772/57492

1. Introduction

Gas chromatography (GC) is a widely applied technique in many branches of science and technology. For over half a century, GC has played a fundamental role in determining how many components and in what proportion they exist in a mixture. However, the ability to establish the nature and chemical structure of these separated and quantified compounds is ambiguous and reduced, and requires a spectroscopic detection system. The most used, is the mass spectrometric detector (MSD), which allows obtaining the "fingerprint" of the molecule, i.e., its mass spectrum. Mass spectra provide information on the molecular weight, elemental composition, if a high resolution mass spectrometer is used, functional groups present, and, in some cases, the geometry and spatial isomerism of the molecule.

2. Gas chromatography

In a gas chromatographic system, the sample to be analyzed may be a liquid solution or a collection of molecules adsorbed on a surface, e.g., the solid-phase microextraction (SPME) system. During the transfer into the GC, the sample is volatilized by rapid exposure to a zone kept at relatively high temperature (200-300°C) and mixed with a stream of carrier gas (Ar, He, N_2, or H_2). The resulting gaseous mixture enters the separation section, a chromatographic column, which in its current version is a fused-silica tubular capillary coated internally with a thin polymer film. Upon their displacement through the column, analyte molecules are partitioned between the gas carrier stream (mobile phase) and the polymer coating (stationary phase), to an extent which depends mainly on their chemical structure. At the end of the separation section, the molecules reach a detection system in which a specific physical property (thermal conductivity) or a physico-chemical process (ionization in a flame, electron capture) gives rise to an electric signal which is proportional to the amount of molecules of the same identity. A data system permits to process these data to produce a graph of the variation of this detector signal with time (chromatogram). Thus, four principal sections are distinguishable in the chromatograph: introduction (injector), separation (chromatographic column), detection, and data handling units. Each section has its own function and its responsibility for the quality of the analysis and the results obtained. The injection system, for example, should ideally transfer the sample to the column quantitatively, without discrimination on molecular weights or volatility, and without chemical alteration (decomposition or isomerization). It is a critical step, especially for quantitative analysis. For correct GC operation, among other conditions, this gateway to the column should remain unpolluted, clean, inert, and leak-free. The main requirement for an analyte in GC is that it should be volatile enough to be present in detectable amounts in the mobile phase. Substances with low vapor pressure will not enter the chromatographic column, will accumulate at the injection system, and may eventually clog its conduits. Very polar, thermolabile, ionic and high-molecular weight compounds are not compatible with regular GC analysis. Depending on the molecular structure of the analyte and the functional groups available, it is possible in some cases to obtain a chemical derivative which has a higher vapor pressure and is therefore more amenable to GC analysis.

One of the most important characteristics of the chromatographic column is its resolution, or the ability to separate components with very similar distribution constants between the mobile and stationary phases (K_D). Chromatographic resolution is a function of many operational parameters. Among them, the nature of the stationary phase, mobile phase, temperature, the size of the column, that is, its length (L), inner diameter (ID) and the thickness of the stationary phase (d_f). As the number of components in the mixture increases, and the structural similarity between its components grows (isomerism), longer columns are required for complete compound separation. Alternatively, for the same purpose, one can employ smaller internal diameter columns. Obviously, increasing the length of the column markedly increases the analysis time. So, as the analysis of polyaromatic hydrocarbons (PAHs) or controlled drugs is regularly accomplished using 30 m long columns, the separation of hydrocarbons in gasoline requires longer, 100 m columns.

2.1. Sample preparation for GC analysis

Sample preparation for GC analysis involves techniques which preferentially isolate volatile and semi-volatile substances and prevent the presence of ionic or high molecular weight species in the mixture to be injected into the GC. These techniques can be divided roughly into three major groups: Distillation, Extraction, and Headspace Methods. Additionally, there are some *sui generis* methods which combine techniques from two different groups, for example, SDE, Simultaneous Distillation – Solvent Extraction [1]. Distillation techniques exploit differences (must be large) in physicochemical properties (volatility or vapor pressure). When applied to vegetal material or to other solid matrix which contains volatile compounds, the distillation can be performed in different ways: (1) Steam distillation (SD), (2) hydrodistillation (HD), or (3) Water-Steam distillation. The resulting extracts or distillates are volatile mixtures suitable to GC or GC-MS analysis, which may only require a drying step (addition of sodium sulfate) prior to injection into the chromatograph. Some distillation processes are designed to

http://dx.doi.org/10.5772/57492

isolate particular substances or fractions at reduced pressure (vacuum) in special columns (molecular distillation) and many operate at industrial scale. However, in general, these techniques are not suitable for studying and isolating compounds at trace levels.

In order to improve extraction efficiency and substantially reduce distillation times, microwave radiation has been actively used in the past 30 years as heat source [2,3]. Microwave-assisted hydrodistillation (MWHD) is a common example of a lab-scale technique for essential oil and other volatile mixture isolation which requires about ¼ of the time employed when conventional heating is used in HD. Microwave-assisted extraction is used mostly for the extraction of solid samples (plant material, soil, tissue, etc.), with water or an organic solvent as the extracting agent [4].

The second family of techniques includes various methods of extraction where the analyte isolation is based on differences in their solubilities in solvent(s) or in the adsorption or absorption on a material such as a microporous solid (activated carbon, silica gel, alumina, molecular sieves, etc.) or a porous polymer (PDMS, Tenax, Porapak, Chromosorb, synthetic resins, etc.).

Liquid-liquid extraction (LLE) in continuous or batch mode is the most used solubility-based extraction method for liquid samples. Extraction selectivity is achieved by a proper choice of solvent (polarity, boiling point, dielectric constant, hydrogen bonding capacity, availability, accessibility, cost, etc.). The extract obtained must often undergo clean-up and concentration processes. However, LLE poses a number of technical problems, among others, emulsion formation, high cost, long extraction times, automation difficulties, toxic solvent disposal, possibility of cross contamination, etc. Despite these difficulties, the number of published works that involve the use of LLE grew during the last decade [5]. Solid-phase extraction (SPE) has emerged during the last 4 decades as an alternative to LLE [6]. It substantially reduces the use of solvents and combines extraction, clean-up and target analyte concentration into a single step. In addition, the method includes the possibility of automation. Another solubility-based extraction technique employs a solvent at temperature and pressure above its critical point. Supercritical fluid extraction (SFE) requires an initially high investment in equipment, but has been used increasingly in areas in which traces of the extraction agent are not desired (natural products, environmental, food, forensic, and in many industrial applications). In this regard, carbon dioxide is the most common fluid employed because when the extract is returned to standard temperature and pressure conditions, CO_2 separates completely in a spontaneous manner because it is a permanent gas under this conditions. SFE selectivity is achieved by varying the operating conditions (temperature and pressure), which modify the supercritical fluid density and in turn, the solubility of the analyte [7]. The use of co-solvents (ethanol, acetone) is another option, particularly, to enhance the extraction of polar substances. For GC-MS analysis, the SFE extract often requires additional cleaning steps to remove fats or pigments, waxes or other high-molecular weight compounds (Figure 1).

Soxhlet extraction is a solubility-based extraction method for solid samples. It´s one of the most traditional extraction methods, used in the analysis of soil, polymers, natural products, etc., in areas such as biochemistry, agricultural sciences, biology, or environmental research. This exhaustive extraction affords mixtures which require clean-up to remove heavy components and other interferences. Long experimental times and high solvent consumption in Soxhlet extraction have led to its replacement in many applications by most efficient methods, such as Accelerated Solvent Extraction (ASE) [8]. In ASE, the use of higher temperature and pressure permits to attain higher selectivity while significantly reducing the time and the amount of solvent required. Extraction, concentration and clean-up are performed in a single step, and the equipment may be coupled in line with GC-MS.

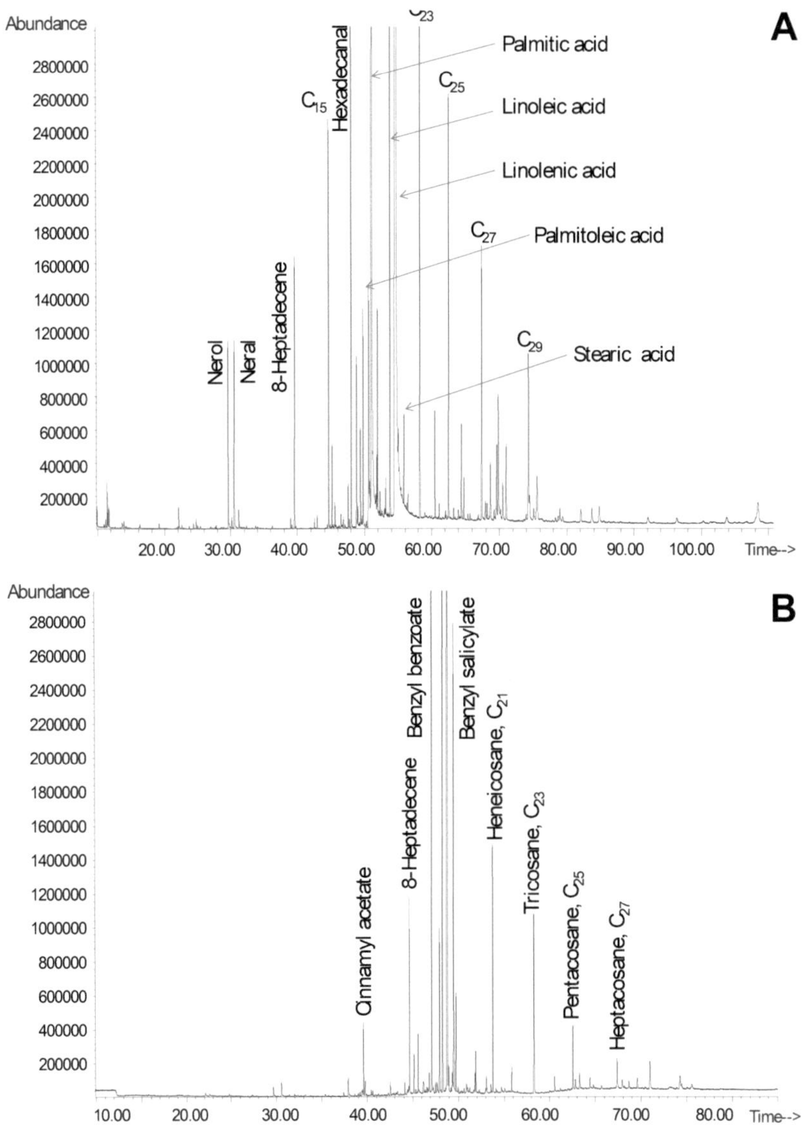

Figure 1. Total ion currents (TICs, chromatograms) of the coffee fresh flower extracts obtained by (A) extraction with hexane (without clean-up) and (B) extraction with supercritical CO_2 followed by clean-up and fat elimination. GC-MS (Electron impact, 70 eV), DB-5MS nonpolar capillary column (60 m x 0.25 mm x 0.25 μm). Split 1:30.

Solid-phase microextraction (SPME) constitutes a fundamental development in sample preparation techniques. In about 20 years of growth, this technique has become the dominant approach to sample preparation for the analysis of volatile substances in biological, environmental, forensic (arson, explosives residue, analysis of drugs in body fluids and tissues) studies, food quality control, or the monitoring of different processes *in vitro* and *in vivo* (breath analysis, fermentation processes, microbiological, pharmacokinetics, etc.) [9,10]. In SPME, the extraction occurs by the mass transfer of analytes from the sample to a polymer coating over a fused-silica fiber. This transfer continues until the chemical potential of each substance in all phases is the same. The fiber is subsequently inserted at the injection port of the chromatograph for direct admission of the analytes into the chromatographic system. SPME sampling with the polymer-coated fiber may take place by exposing it to the headspace above the sample (liquid or solid), or by immersion of the fiber in the sample (liquid), directly, or within a protective membrane. The SPME sampling process is amenable to automation and operation coupled to a GC. One of the most interesting applications of SPME is the target-analyte extraction accompanied by its derivatization on the fiber, which has been previously loaded with a reagent (Figure 2) [11].

The use of the vapor phase above the sample (headspace) as a means for performing extraction processes has great advantages from the analytical point of view, because it eliminates many interferences, mostly the solvent peak, which can impede the detection of volatile compounds. In the study of natural products, direct headspace sampling has allowed the study of the volatile fraction around whole plants or parts thereof (Figure 3) [12]. Classical approaches to monitor the headspace above a sample may be divided into static and dynamic versions. The volatile fractions thus collected are suitable for GC- MS analysis, as they are free of interference from nonvolatile very polar (ionic) or high - molecular weight matrix components. Static headspace consists of the direct transfer of a headspace volume to the GC injection port. This is the sampling method regularly used in the analysis of BTEX (benzene, toluene, ethylbenzene, and xylenes) in water, quantification of trihalomethanes in drinking water, ethanol, volatile compounds in paints, determination of residual solvents in polymers and pharmaceuticals, hydrocarbons in soil, analysis of beverages and perfumes, among many other applications [13]. When the target-analyte concentration in the sample is very low (ppt, ppb), enrichment (pre-concentration) is required, and in these cases, dynamic headspace is the preferred sampling mode. Basically, by purging the vapor phase with an inert gas, the analytes are collected in a solvent or are "frozen " (cryo-trapping) or are ad(b)sorpted in a sorbent (purge and trap, P & T), and later on are recovered by the action of temperature (thermal desorption) [14] or are eluted with solvent.

Sorption-based sampling techniques, which involve aspects of both static and dynamic headspace, have been developed in the past 2-3 decades. The great development experienced by the SPME technique has served to identify, through its many applications, some areas for improvement. This has generated a collection of new methods of micro-scale sampling, that achieve better concentration of the analytes of interest and in several cases, the use of a larger volume of sorbent phase (polymer, normally) permits the collection of larger amounts of analyte than SPME. In the inside needle capillary adsorption trap (INCAT), a polymer layer

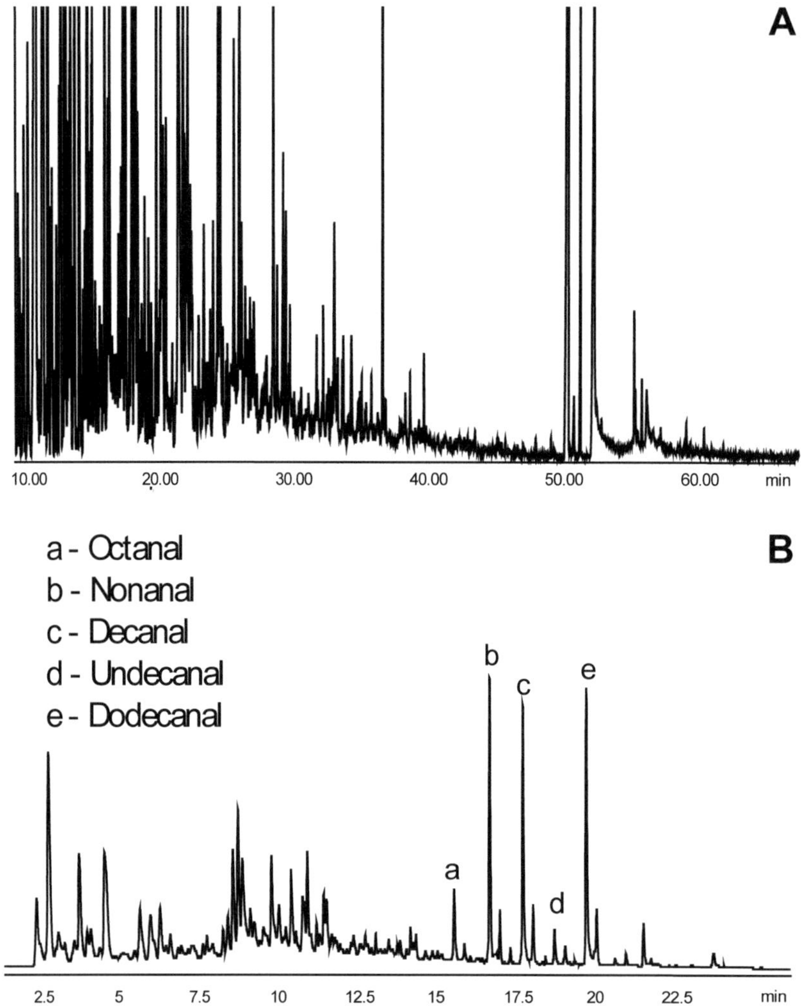

Figure 2. Total ion current (TIC, chromatogram) of the coffee roasted beans volatile fraction isolated by (A) P&T (purging with nitrogen and trapping with dichloromethane). GC-MS (Electron impact, 70 eV), DB-5MS nonpolar capillary column (60 m x 0.25 mm x 0.25 μm), split 1:30; and (B) Headspace-SPME with simultaneous aldehyde on-fiber derivatization with pentafluorophenylhydrazine (PFPH). Temperature: 50°C and exposition time: 30 min. GC-ECD, DB-5MS nonpolar capillary column (30 m x 0.25 mm x 0.25 μm). Split 1:30.

coats the inside of the needle of a syringe, or the inside is filled with activated carbon [15]. Headspace solid-phase dynamic extraction (HS-SPDE) is a sampling technique in which the needle of a hermetic gas sampling syringe is coated with absorbent polymer (PDMS) on which the analytes are deposited when the syringe is actuated repeatedly over the sample [16]. The number of suction and discharge cycles determines the amount of accumulated analyte. Stir bar solid-phase extraction (SBSE) and tape sorption extraction (STE) are additional examples of sampling techniques using higher absorbent phase volume. In the first one, the magnetic stir bar is coated with absorbent polymer. Upon exposure to the solution containing the

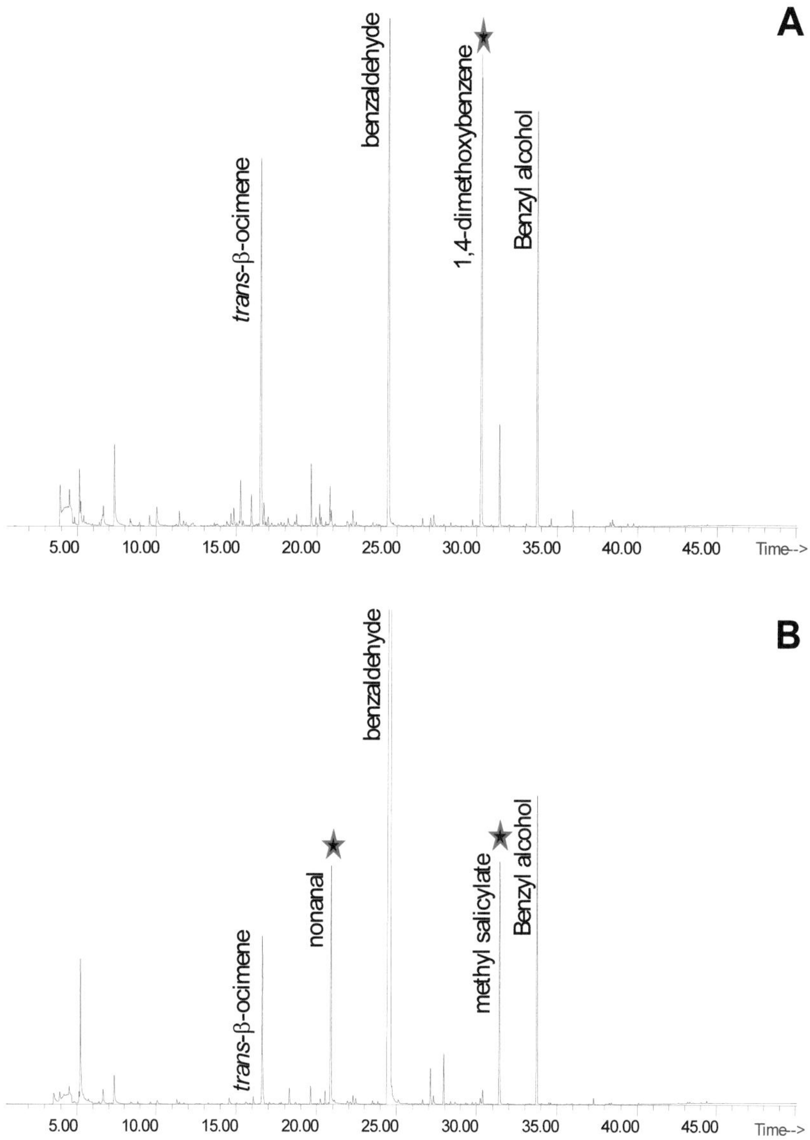

Figure 3. Total ion currents (TICs, chromatograms) of the passion fruit flowers (*Passiflora edulis*) volatile fractions obtained by headspace- SPME from different flower parts: (A) male stamens and (B) female carpels. Notice the composition differences which are very important in the pollination processes. CAR/PDMS-fibre coating. Headspace-SPME exposition time: 30 min. GC-MS (Electron impact, 70 eV), DB-5MS nonpolar capillary column (60 m x 0.25 mm x 0.25 µm). Splitless injection mode.

analytes of interest, the stir bar is subjected to thermal desorption for transfer of the analytes to the chromatograph [17]. In STE, a thin strip of PDMS is placed for a fixed time in direct contact with the plant surface, the skin, or the object to be sampled. Analytes are subsequently transferred to the chromatograph by thermal desorption [18].

2.2. Sample introduction systems

Another absolutely essential parameter to determine how many components in a mixture are detected is the injection mode of the mixture into the GC. The main objective of an injection system, or simply the injector or the injection port, is to transfer the sample to the chromatographic column in a rapid and quantitatively reproducible manner. Ideally, this critical step in GC, has to meet a number of conditions. These include: (1) The sample must enter the column in a chromatographic band as thin as possible, to reduce peak broadening; (2) the percentage composition or relationship among the components of the mixture analyzed should be the same before and after injection; that is, there should be no discrimination by molecular weight, or volatility at the injection port; (3) during their transfer to the column, components should not react with each other or undergo chemical change, e.g., isomerization, hydrolysis, polymerization, etc. or become adsorbed onto the injector body. The sample must be transferred quantitatively without alterations. The sample injection must be precise, accurate, reproducible and predictable, quantitative and without discrimination of any kind. The injection mode is selected according to the following criteria: (1) type of chromatographic column (packed column, mega–bore, capillary), (2) the chemical nature of the analytes (molecular weight, volatility, polarity, temperature stability, reactivity, etc.) and (3) the purpose of the analysis (qualitative or quantitative, to detect traces or, conversely, to measure high concentrations of some components, etc.).

Here are some considerations on the most common injection ports for capillary columns. The "universal" injector does not exist. For capillary columns two main types of injectors are available: vaporization and on-column (direct injection into the column without prior sample volatilization) injectors. There are also devices that can be connected in line with the GC injection port, for example, headspace sampler, pyrolysis, a purge and trap system, a thermal desorption system, a switching valve to transfer gas samples, etc. The injection of the sample can occur either manually or via an autosampler. In the vaporization injector, the sample (1 – 2 μL) crosses a septum to enter a glass cylinder which acts as the liner of a chamber which is heated independently. Upon exposure to the high liner temperature (> 200 °C), the sample vaporizes quickly and the resulting gaseous mixture is swept towards the column by the carrier gas. Conventional vaporization injectors can operate in two primary modes, i.e., split (sample is divided at the column inlet) and splitless (without division of the sample) [19]. In a standard split/splitless injection port, the gas flow may transport all of the injected sample to the column (splitless mode) or only a fraction of it (split mode). Concentrated samples are injected in the split mode so that only a small aliquot enters the column (Figure 4). On the other hand, when analyzing trace level compounds, for example, with environmental or forensic samples, complete transfer of the sample to the column, without division, is required (splitless mode). This mode involves slowly transferring (ca. 1 mL/min) the sample to the column through the hot injection port, which may lead to thermal decomposition and/or isomerization of some components of the mixture. A special pulsed splitless mode has been devised to enable the analysis of thermally labile compounds at trace levels, which is performed by increasing pressure on the top of the column during the injection, while the sample is transferred to the column. In this case, the residence time of the sample in the hot region (250-300 °C) of the injection port is dramatically reduced and so is the probability of the thermal decomposition of some components (Figure 5). A mode of complete sample transfer to the column, which avoids component discrimination by their volatility or molecular weight, is the on-column injection, i.e., applying the sample directly to the column head. Very detailed work on the design and application of various methods of sample injection to GC equipment has been developed by Dr. Konrad Grob [19,20].

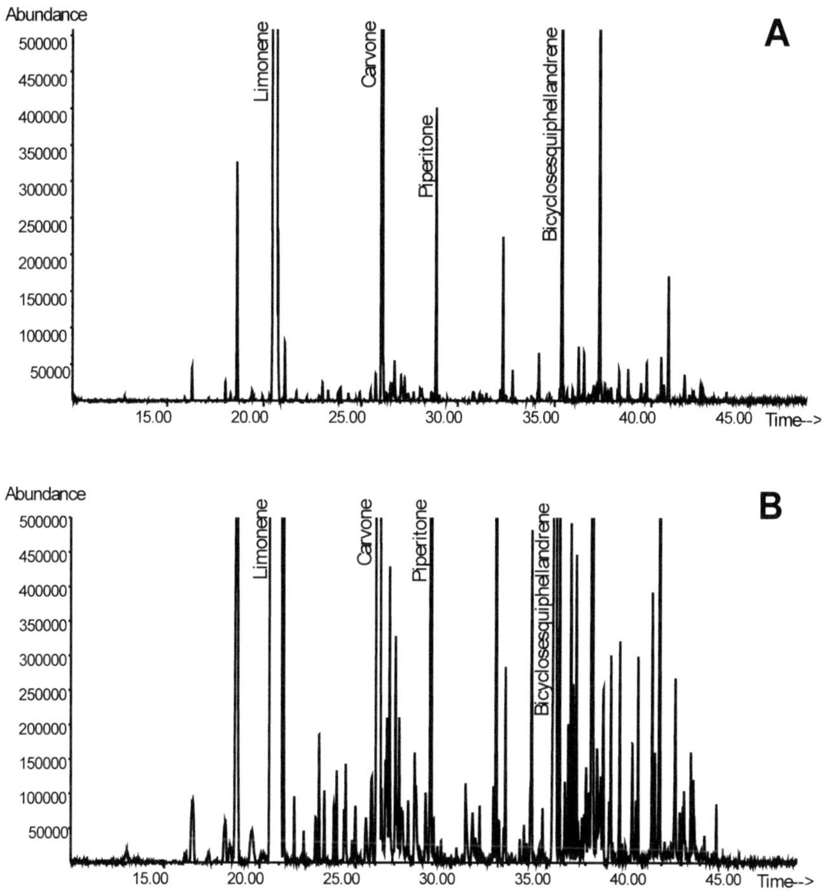

Figure 4. Total ion currents (TICs, chromatograms) of the *Lippia alba* (Verbenaceae family) essential oil obtained by hydrodistillation. A. 1:30 Split injection (ratio, 1:30, Inj. vol.1 μL) and B. Splitless injection mode (Inj. vol.1 μL). Almost twice the number of compounds were registered in the chromatogram obtained by splitless injection (ca. 90) in comparison to those registered in split-mode injection (>0,05%). GC-MS (Electron impact, 70 eV), DB-1MS nonpolar capillary column (60 m x 0.25 mm x 0.25 μm).

Different configurations of the liner offer efficiency and uniformity of the mixing process. A specific liner configuration is recommended for each type of injection, information which can be found in the catalogs of different manufacturers of these accessories. The inlet temperature should be high enough to ensure complete evaporation of all analytes, but not so high that

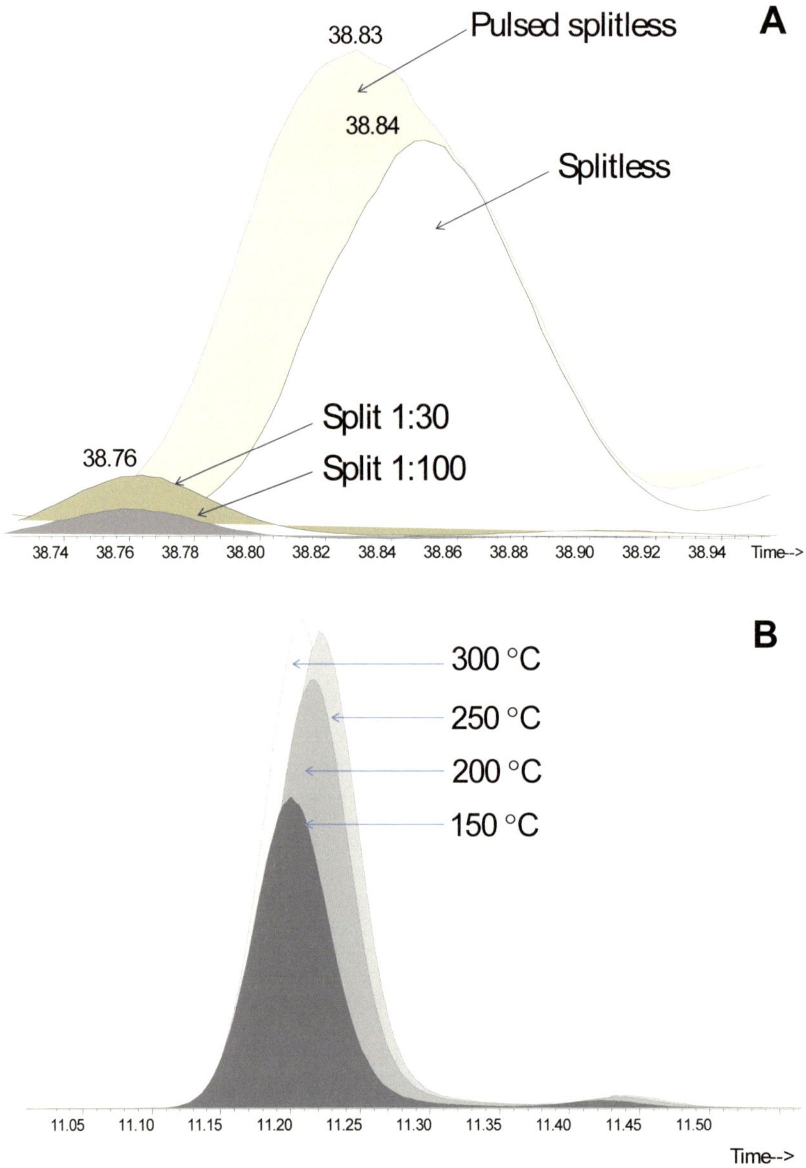

Figure 5. Chromatographic peak of ethyl benzoate obtained by GC-MS (Electron impact, 70 eV), on the DB-5MS non-polar capillary column (60 m x 0.25 mm x 0.25 µm). The sample was injected in different modes: split, splitless and pulsed splitless. Notice the retention times shift and the change of chromatographic areas. B. The influence of injection port temperature on the chromatographic peak area of limonene (monoterpene, $C_{10}H_{16}$). DB-5MS column (60 m x 0.25 mm x 0.25 µm). 1:30 Split ratio.

causes thermal decomposition. Too low temperatures will produce incomplete volatilization, a loss of analytes and a slow entry of volatilized substances to the column, causing chromatographic peak broadening. A potential problem is the systematic contamination of the injector and the development of a "memory effect". More commonly, the temperatures used in conventional vaporizing injectors range from 200 to 320 °C, the temperature of 250 °C being quite common in the practice of GC [21-24]. It is important to note that when the liquid sample is vaporized, it expands many times with respect to its initial volume. Expansion volumes depend on: (1) the type of solvent, (2) inlet temperature and (3) inlet pressure of the carrier gas at the top of the column [23]. Before injection, all these factors must be taken into account, i.e., what amount of sample to be injected, in which solvent and under what conditions (injection temperature and inlet pressure in the column). The flashback, i.e., the "return" of the sample vapor by the purge gas line, may occur when the expanded volume of the injected sample volume exceeds that of the liner (ca. 0.5- 1 mL). Excess vapors can escape and pressure is created not only by the carrier gas line (flashback), but also by the septum purge line; the more volatile components of the sample can be "lost" and this will affect the accuracy and precision of the GC analysis.

The injection port is a critical site in any chromatographic system, since it may be the cause of contamination (high noise) and the source of so-called ghost peaks, whose origin may be related to recondensation of the less volatile substances in the bottom of the septum or the carrier gas transfer line from flashback problems [24]. Injection efficiency, optimum speed and consistency of the mixture entering the column will determine the width of the chromatographic peak, its symmetry, and, ultimately, efficient and reproducible separation and quantification. When choosing the injection technique, the following variables are chosen and optimized: (1) the injection mode (split, splitless, on-column or programmed temperature vaporizer, PTV), (2) the volume of the sample to be injected, (3) injector temperature, (4) the type of liner (shape, volume, packaging), (5) the initial temperature of the column (more critical for injection in splitless mode), (6) the form of injection, manually or automatically, (7) the rate at which the septum is pierced or the needle is removed from the injection port (important parameters for decreasing analyte discrimination based on volatility), (8) the solvent and the number of washes of the syringe with different solvents, (9) the type of septum (material, density, thermal stability, bleeding), for instance, the conventional septum or Merlin Microseal™, which is highly recommended for GC-MS operations.

2.3. Separation procedure

The chromatographic column is the heart of the GC. There are two types of columns, packed and open tubular, often called capillaries (when the ID is equal to or less than 0.32 mm). In packed columns, solid particles (silica, alumina, molecular sieves, porous synthetic polymers, etc.) serve as support or are embedded in a liquid (squalene, polyethylene glycol, etc.), which acts as a stationary phase as means of separating components within the GC. Because the gas solubility in liquids is very low, solid phases are particularly suitable for the separation of gaseous components, such as permanent gases, natural and refinery gas, volatile aldehydes and alcohols, car exhaust, etc. The use of packed columns is fairly limited today, because open tubular columns have much higher separation efficiencies. Current gas analysis uses porous layer open tubular columns (PLOT), a type of capillary columns, where analyte adsorption is combined with the high separation efficiency provided by capillary columns [21].

Column efficiency is expressed by the so-called number of theoretical plates, N, or by means of the height equivalent to a theoretical plate, HETP [22]. The higher is the number of N and lower the value of HETP, the higher the resolution (efficiency) of the column. As indicated by its name, open tubular columns have an opening, an internal space, which allows the carrier gas to move freely without having to penetrate or pass through, with strength, the porous support. This latter process, the tortuous pass, accompanied by the phenomenon of multiple paths, is typical for packed columns and is the main cause of their much lower efficiency compared with open tubular columns [22]. The GC column selection process should take into account the following parameters and properties: (1) phase type (nature, polarity), (2) dimensions (length, L, and inner diameter, ID) and (3) phase thickness. Stationary phase polarity is chosen according to the polarity of the analytes of the mixture, following the general principle of "like dissolves like". But often, it is difficult to find a “universal” stationary phase, since in many samples both polar and nonpolar compounds can be found, with diverse volatility (Figure 6). In nonpolar columns (polydimethylsiloxanes), analyte elution happens according to their volatility, i.e., in increasing order of their boiling points. An example is the technique of simulated distillation for the analysis of petroleum fractions [25]. In polar columns (e.g., polyethyleneglycols), the predominant factors for analyte separation are its polarity (the dipole moment, μ) and the strength of its interaction with the stationary phase (dipole, hydrogen bonding). Wall-coated open tubular columns (WCOT) are the most commonly used in capillary GC and their stationary phases can be divided, roughly, into two great classes: (1) substituted siloxane-type polymers with a wide range of polarities and (2) poly(ethyleneglycol) (PEG), which are highly polar. The polysiloxanes may possess different substituents which provide to them some polarizability. The polysiloxane phases range from nonpolar 100% poly(dimethylsiloxane) slightly polar 5-95% -phenyl poly(methylsiloxane), to polar and strongly polar phases containing trifluoropropyl and cyanopropyl substituents, respectively. Modern GC columns have cross-linked and immobilized (chemically bonded to the column wall) polymer stationary phases. This provides uniformity, chemical and thermal stability, and reproducibility of results during extended use. Chiral stationary phases for separating optical isomers give rise to a separate family of columns. Among these, good performance has been achieved by the presence of cyclodextrins within the stationary phase. Enantiomers appear in the chromatogram with different retention times due to the different strength of their association with the cyclodextrins when these adducts involve the non-polar cyclodextrin´s inner cavity [26].

Capillary columns are divided according to their length, L, into: (1) short (5-15 m), (2) medium (20-30 m) and (3) large (50-100 m). Perhaps the length L = 30 m is the most common column used for many analyses (drugs, pesticides, polyaromatic hydrocarbons, etc.) [21,22]. With increasing column length the number of theoretical plates (N) and resolution also increase, but the duration of the chromatographic run is greatly increased. Long columns are required for the analysis of complex, multicomponent (> 50 substances) samples, for example, gasoline, perfumes, essential oils, aromas, VOCs, etc. [22,23], or when there are isomers in the mixture which have similar boiling points or dipole moments and, consequently, their distribution constants, K_D, are very close. The longer the columns are, the greater the pressure drop across them (the pressure difference between inlet and outlet of the column). Also, due to their greater

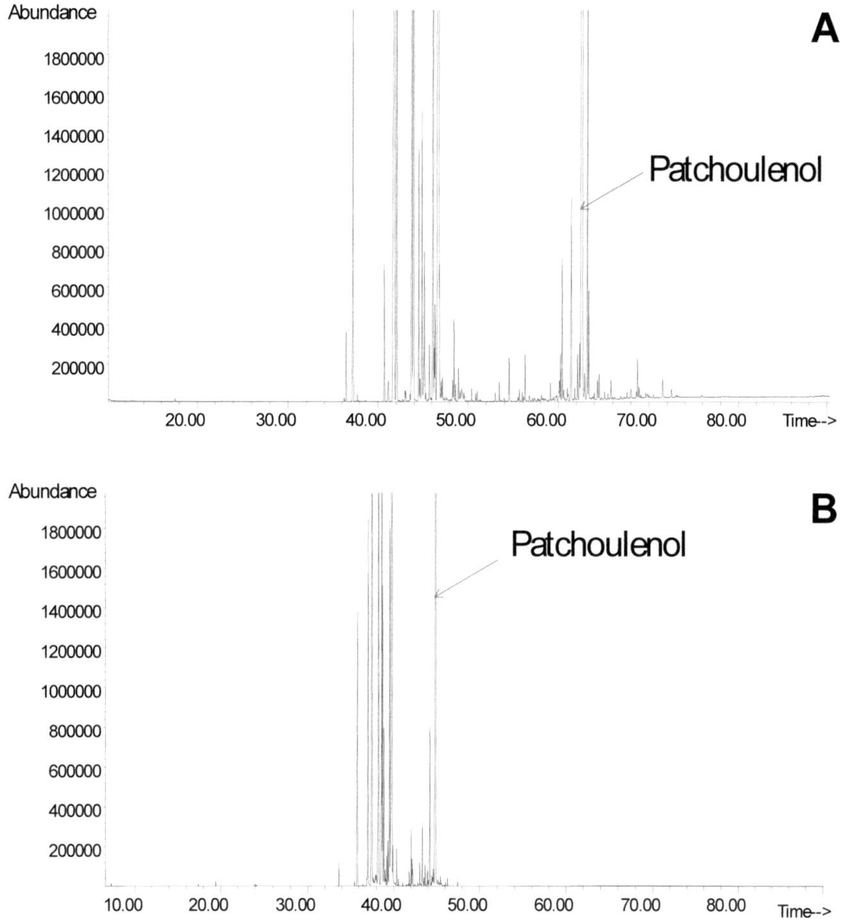

Figure 6. Comparison of the total ion currents (TICs, chromatograms) of the patchouli (*Pogostemon cablin*) essential oil obtained by hydrodistillation. The complex mixture of sesquiterpene hydrocarbons and their oxygenated derivatives (alcohols and aldehydes) was analyzed in two capillary columns of the same dimensions (60 m x 0.25 mm x 0.25 m), but different stationary phases: (A) DB-WAX and (B) DB-5MS. GC-MS (Electron impact, 70 eV), Split 1:30. The same oven temperature program was used with both columns: 50 °C (5 min), 4 °C min^{-1} to 250 °C (15 min).

thermal inertia, the heating rate of these columns must be slower (2-6 °C min^{-1}), in order to achieve reproducible change in temperature. The analysis of polyaromatic hydrocarbons (PAHs) can be successfully accomplished in 30 m-long columns, whereas the separation of hydrocarbons in gasoline requires longer 100 m columns. The main drawbacks of long columns are extended analysis time and higher cost, but very high sensitivity and resolution can be achieved. A compromise has to be found among resolution, sensitivity and time (cost) of analysis. A significant alternative to increase the resolution of the column and shorten analysis time is to reduce the column internal diameter, which in capillary columns most commonly is 0.25 or 0.32 mm. When the column length is doubled, the resolution is also doubled, but when ID is reduced by half, the resolution increases almost fourfold. Decreasing I.D. leads to a marked reduction in analysis time. This led to the development of a new branch in GC, named fast chromatography, which has found very interesting applications [27-29]. The columns employed are thin or ultra-thin (0.1 or 0.05 mm), but shorter (1-5 m) than in conventional GC [29]. The "fast" columns also require a detector with very fast response, particularly in the case of comprehensive GCxGC; in the fast -GC-MS configuration, the most effective mass analyzer is the time-of-flight (TOF) [30]. The main shortcoming of fast GC is its low sample capacity, which for many analyses, for example, environmental or forensic, can become a serious limitation. Inlet pressures in the overhead GC column increase greatly as the ID is reduced. For example, 30 m columns (100 °C) with IDs of 0.32 and 0.10 mm require carrier gas (helium) pressures of 9.5 and 114 psi, respectively. Roughly, the resolution of the column increases with its length and with the decrease of ID and phase thickness, while the analysis time increases with the rise of the column ID and of the thickness of the stationary phase and, above all, with the length of the column. The main tasks after correctly selecting the column should focus on the installation of the column and its conditioning, maintenance and precautions to be taken during operation. The latter is related to the care of the GC column, which includes not exceeding the permitted temperature limits, preventing the entry of oxygen, water, mineral salts, acids or bases or high molecular weight compounds (interference), maintaining a hermetically sealed GC system, devoid of leakage. With time, gaskets, seals, traps (for moisture, hydrocarbons or oxygen) wear out or saturate and should be replaced.

For capillary columns, the most recommended carrier gases are helium or hydrogen; an average linear velocity of 25-35 cm/s allows obtaining the required separation efficiency. Since the gas viscosity varies with temperature, the average linear velocity of the carrier gas will decrease during the chromatographic run as oven temperature increases. For substances with high distribution constants (K_D), this reduction in the flow rate of the carrier gas impinge upon the shape of their chromatographic peaks (asymmetry, tailing) and increases the retention times and the overall analysis time. Hence, special devices have been introduced, the electronic pressure controls, which, among other functions, allow maintaining constant flow in the column during the chromatographic run so that carrier gas pressure input to the system is changed with increasing oven temperature, in a synchronized, electronically governed manner.

2.4. Multidimensional and comprehensive separation systems

Essential oils, fragrant liquids consisting of volatile secondary metabolites isolated by distillation of aromatic plants are multicomponent mixtures, which may contain up to 300 substances [31]. The essential oil is an excellent model for gas chromatography, as its constituents possess different functional groups; among them are hydrocarbons, alcohols, aldehydes, ethers and esters, acids, and others. Most of these substances belong to the family of terpenoids (C10 and C15), but also include phenolic derivatives. The concentration of the constituents in the essential oils can vary over a wide range, from μg/kg or mg/kg to g/kg. Many components of the oils have isomers. This leads to two problems in reaching the analytical goals, namely: (1) total separation of all components, and (2) the correct assignment of their structures. For some substances present in the essential oils very close retention times are observed, at least in one of the columns (non-polar or polar); in addition, some constituents possess very similar mass spectra. Complete separation of the components of essential oils can be achieved with complete gas chromatography (comprehensive GCxGC), which has been applied very successfully to such mixtures [32]. Multidimensional chromatography allows separating in a second column the peaks of partially or completely co-eluted substances, through the operation of "heart -cutting" using switching pneumatic valves (nowadays micro-fluidics technology) between the two columns and diverting part of the effluent from the first to the second column. This technique has played an important role in the development of methods for complex mixtures separation. Multidimensional chromatography requires at least two detectors and some configurations can have three columns in the same or in separate chromatographic ovens.

Along with this and other conventional approaches, nowadays there is a modern solution, although relatively inaccessible to many laboratories, due to its cost, for the separation of multicomponent mixtures. It is called complete, total, or more commonly, comprehensive chromatography, GCxGC. It uses two columns connected by a modulator. In contrast to conventional multi-dimensional gas chromatography, GC x GC requires a single detector and both columns can be in the same or in two separate ovens. There are different types of modulators, e.g., thermal rotary modulator (sweeper), "jet" cryogenic system, valve modulator, and longitudinal cryogenic modulator, among others [32]. The eluent from the first column is split into very small slices by means of the modulator, which transfers them one after another, in a row, from the first into the second column. The first column (1D) is a conventional column, with length of 25 or 30 m, and the second column (2D) is a fast chromatography, a very short micro-bore column. Stationary phases in both columns are "orthogonal", i.e., if the first one is non-polar, the second column is polar, and vice versa. Modulation time, i.e., the transfer of a tiny portion of the first column eluent to the second, must be very short and commensurate, but not longer, with the elution time of a component in the second column. Thus, the latter is very short and very thin and allows separating the components in a few seconds. Since the second column is connected to the detector (MSD, FID or μ-ECD), it should have high reading and signal processing speeds. In most cases, the time-of-flight mass detector (TOF) is the best, although still very expensive option [32].

2.5. Overview of GC detectors

The detection system differentiates the analyte molecules from the mobile phase, which is transparent to the detector. Detector response (signal) is based on measuring a physical property of the flow of analyte molecules (ionic current, thermal conductivity, fluorescence, refractive index, photon emission, electron capture, etc.). The signal should be proportional to the amount of analyte that emerges from the column, thus establishing an interdependent relationship and carrying out a quantitative analysis, which is an essential part of a chromatographic determination and leads to the answer on how many components and in what proportion they are present in a mixture. Common detectors used in gas chromatography are classified into (1) universal detectors, e.g., thermal conductivity detector (TCD), mass selective detector (MSD) operated in full scan mode, or infrared detector; (2) selective detectors, e.g., nitrogen-phosphorus detector (NPD), electron capture detector (ECD) or flame photometric detector (FPD), among others, and (3) specific detectors, e.g., MSD operated in selected ion monitoring (SIM) mode, thermal energy analyzer (TEA), or atomic emission detector (AED). The flame ionization detector (FID) could be considered "near- universal" since only water and permanent gases are transparent. There is no well-defined boundary between the specific and selective detectors [33]. Specific detectors are actually highly selective detection systems, which can obtain a response to one particular compound, an analyte or target of interest present in a complex mixture. For example, in a urine sample, detect and quantify testosterone, which is one of the analytes to be determined quantitatively for doping control in sports. The selectivity of a detector response is related to a substance that has a specific atom, for example, nitrogen, phosphorus (NPD) or sulfur (FPD) or a functional group, e.g., electronegative group(s) (ECD) or unsaturated groups and aromatic rings (PID, photoionization detector) or the substances, which have a common structural fragment, for example, a phenyl, benzyl or an acyl radical, etc. (MSD operated in SIM mode; Figure 7).

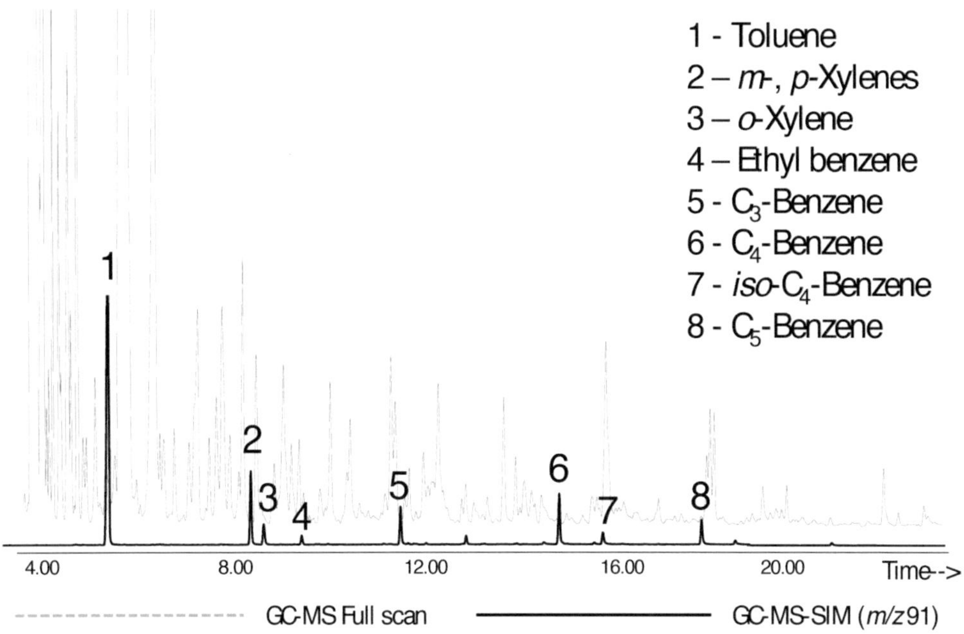

Figure 7. Selective mass-spectrometry detection of alkyl benzenes present in the volatile fraction isolated by P&T from roasted coffee beans. GC-MS-SIM (Electron impact, 70 eV, selected monitoring of *m/z* 91 ions, characteristic for alkyl benzenes). DB-5MS nonpolar capillary column (60 m x 0.25 mm x 0.25 µm). Split ratio 1:30.

All detectors are distinguished by their sensitivity, minimum detection and quantitation levels (in the case of MSD, also by the minimum identification level), linearity, sensitivity to changes in gas flow, temperature or pressure. The sensors have different noise levels, volumes, are distinguished by a higher or lower sensitivity and sophistication, for simplicity in operation, cost, and other typical features to be taken into account when choosing and operating a GC detector. Because the volumes of GC detectors are larger than the capillary column attached to them, the dead volume formed could generate chromatographic peak broadening and affect resolution. This difference in volumes is offset by an auxiliary gas (make-up), usually nitrogen, whose flow is 20-40 mL/min, depending on the detector type. High make-up flows lead to more symmetrical and thin chromatographic peaks, but compromise sensitivity. The trend in modern GC instruments is to reduce the volume of the detector. Today, there are micro versions of several detectors, for example, μ-ECD or μ-TCD. Most of the GC detectors are destructive detectors. These include, for example, ionization detectors (FID, NPD, FPD, MSD), whose response is sensitive to the change in mass of the analyte, while the response of non-destructive detectors (TCD, infrared detector, ECD, PID) is sensitive to the change of the analyte concentration. That is, when operating the TCD or ECD it is important to maintain the flow of gases (carrier gas and make-up or auxiliary) constant. Nondestructive detectors can precede in series a destructive detector; e.g., in tandem TCD followed by FID or NPD.

The major limitation posed by conventional detectors (FID, TCD, ECD, NPD) is the ambiguity to uniquely identify an unknown substance. Absolute retention times (t_R) or relative to an internal standard (t_{RR}) and certified standards, are used for the presumptive identification of a compound. This is a screening analysis. Confirmatory identification of a compound in a complex mixture, analyzed by GC, necessarily requires obtaining the "fingerprint", i.e., spectrum, of the compound. The mass spectrum (MS) has information about a unique combination of charged fragments (ions) generated during the dissociation or fragmentation of the previously ionized molecule. The complementarity of chromatographic analysis (screening) with confirmatory spectral data is achieved using the combination of GC-MSD. A GC-MSD analysis provides information on (1) retention times, (2) chromatographic areas and (3) mass spectra of each component in a mixture, obtained by its ionization with electron impact (EI) at 70 eV. The value of 70 eV was established as a standard to obtain the MS, which are also part of the databases and commercial spectral libraries, because for most organic compounds, the highest ionization efficiency and reproducibility are achieved with bombarding electrons of this energy. The ionization energy (potential) of organic molecules varies from 7 to 14 eV. During the collision with electrons of 70 eV, energy is transferred to the molecules to ionize them and form the molecular ion, M^{+}, but also to produce fragmentation of those excited M^{+*} ions, which possess sufficient internal energy to break a chemical bond. The spectra obtained with lower energy electrons (15-20 eV) are called low voltage spectra, but they are rarely used in the GC- MSD method, because they often cannot achieve the sensitivity required for the analysis.

The coupled technique GC-MSD is the most used tandem combination in instrumental analytical chemistry and is applied to investigations in forensic, environmental, natural products, foods, flavors and many other areas [34]. Among the chromatographic spectroscopic detectors (AED, IR or MSD), useful for the determination of the chemical structure or elemental composition of analytes, the mass spectrometer is the most widely used today thanks to its sensitivity, operability and, above all, to a large amount of *sui generis* structural information that it can provide [35].

2.6. Mass selective detector

A mass spectrometer attached to the gas chromatograph is often referred to as "mass detector" or "mass selective detector" (MSD). The MSD consists of an ionization chamber which in the large majority of situations uses electron ionization or electron impact (El) to provide the energy to the analytes which could fragment and generate the ions to be detected. Chemical ionization (Cl) of positive ions or negative ions, is a complementary, mild ionization method. CI is used as an alternative when no molecular ions are registered in the mass spectra obtained by El, since fragmentation is excessively intense, due to the high lability of molecular ions, $M^{+\bullet}$ (extremely short lifetime, $< 10^{-6}$ s) [36]. Ions formed in the ionization chamber are removed therefrom by a series of electrodes that collimate (focus) and accelerate and direct them to a mass analyzer. The potential energy created by the accelerating field (E) is converted into kinetic energy. The analyzer separates ions according to their *m/z* ratio. Several types of mass analyzers may be used in the GC-MS equipment. Quadrupole (Q) and ion trap (IT) are the most common; in recent years, there has been an increase in the use of time-of-flight analyzers (TOF); lately, resonance ion-cyclotron analyzers with Fourier transform (FT-ICR) and orbitrap have become commercially available. The use of magnetic deflection mass analyzers is much less frequent [34,37,38].

The charm –from an analytical standpoint-, of a mass selective detector, is its ability to operate in three modes of data acquisition, namely: universal, selective and specific. When a scan is complete (the MSD detector functions as a universal detector), the mass spectra obtained for all mixture components are the basis for recognizing or identifying them. The mass scanning rate depends on the type of analyzer used (quadrupole, ion trap, time-of-flight or magnetic sector) and the mass range to cover. The mass range for the entire sweep is established in accordance with the nature of the sample, that is, the range of molecular weights of its components. The low end of the scanned mass range for aliphatic compounds, alcohols, amines, etc., can be set to *m/z* 30-40 (lower masses are not recommended, as they correspond to background signals); the minimum mass for aromatics can be set to *m/z* 50. The upper end of the scan range should correspond to the molecular weight of the heaviest substance present in the mixture, plus 40 - 50 units. If the mass range is chosen poorly, for example, if for the GC-MSD analysis of low molecular weight, highly volatile compounds (molecular weight less than 150 Da), a very wide mass range is set, e.g., *m/z* 30-550, the number of spectra obtained per unit of time will be small, which affects the quality, reproducibility and reliability of the analytical data [34-36]. On the other hand, high scan speeds demand fast response from the accompanying hardware, to avoid loss of information.

Selected ion monitoring, SIM, is the selective mode of operation of the MSD. Instead of scanning a mass range, only a discrete set of *m/z* values is monitored. This permits to obtain a much higher signal accumulation per unit time, which means that the sensitivity is greater (30-100 times) than when full scan is used. GC-MS-SIM analysis of an extract or a complex mixture permits the selective detection of homologous molecules, isomers, or structural derivatives, when the selected ion is a common feature in their mass spectra. However, the detection becomes specific, highly selective, when the monitored ions constitute a unique combination which is found only in the mass spectrum of the target analyte.

http://dx.doi.org/10.5772/57492

Besides its use in performing quantitative analysis with higher selectivity and sensitivity, the SIM mode helps to reduce the possibilities of false positives in substance identification. Presumptive identifications are based solely on matching retention times of target analytes and standard compounds, but this result should be confirmed by GC-MSD-SIM, when compounds of interest are in the mixture at trace levels [39].

To setup a GC-MSD analysis in SIM mode, generally three characteristic "diagnostic" ions are chosen (one for quantification and other ones as "qualifiers"). The ion selection criteria attend the following recommendations: 1) the ion´s abundance should be higher than 30%; 2) the mass of the selected ion should be high, preferably, because low-mass ions are common to many substances; 3) the selected ion should be structurally representative of the molecule. For example, it could be a molecular ion, or a genetically-related fragment; 4) the selected ion should not coincide with those from background (*m/z* 17, 18, 28, 32, 40, 43, 44), stationary phase bleeding (*e.g.,*, *m/z* 73, 147, 207, 281, 355), thermal degradation of the septum, or plasticizers (*m/z* 149). The partial ion currents of the selected ions are measured in a retention time window in which the standard substance elutes (e.g., $t_R \pm 0.5$ min). Then the sample is analyzed under the same GC-MS-SIM operational conditions used with the standard substance. When there are several substances of interest, the same procedure is applied to each one and the selected ions are scanned only within the retention time window of its corresponding standard substance. The identity of the target analyte in the sample can be confirmed only when the following 3 criteria are satisfied: 1) the retention times of the analyte and the standard substance match; 2) the S/N ratio is higher than 1:3 – 1:10; 3) the intensity ratios of the selected ions are the same in the spectra of the analyte and of the standard substance. If the retention times are the same (a necessary, but not sufficient condition), but the intensity ratios of the selected ions in the spectra of the target compound and of the standard do not match, that is, are not identical, differ by more than 15%, the situation of a " false positive" is observed, i.e., the alleged presence of the target analyte in the sample is not confirmed [40].

3. Gas chromatography-mass spectrometry

The need to unequivocally identify the components of complex mixtures was the motivation for the development of different instrumental coupling techniques (tandem), including the widely and successfully used (with volatilizable substances), gas chromatography coupled with mass spectrometry (MS). GC-MS is an extremely favorable, synergistic union, as the compounds susceptible to be analyzed by GC (low-molecular weight, medium or low polarity, in ppb-ppm concentration) are also compatible with the MS requirements. Besides both analyses proceed in the same aggregation state (vapor phase). The only "conflict" (short-term and already resolved) between GC and MS, were the different working pressures, i.e., atmospheric at the GC column exit and low (10^{-5} - 10^{-6} Torr) in the ionization chamber, respectively. This drawback was overcome by technically introducing an efficient vacuum pump (turbomolecular and gas-jet pumps) and, above all due to the introduction of gas chromatography capillary columns (internal diameter 0.18 to 0.32 mm id, traditionally used in GC- MS), which are inserted directly into the ionization chamber of a mass detector [34,40].

3.1. Ionization modes

The essence of a mass spectrometric method revolves around the process of ionization of the molecule, with or without subsequent cleavage or fragmentation. In most cases the ionization of the molecule is dissociative. Its mechanisms can be various, e.g., subtraction or addition of an electron, protonation or deprotonation, nucleophilic, electrophilic addition or subtraction and cluster formation, among some other processes leading to ion formation. The ionization of a molecule is an energy consuming process, which can be supplied by accelerated or thermal electrons (electron impact or electron capture), by photons (photoionization, corona discharge, laser beam), by atoms or ions accelerated by a high electrostatic field gradient or thermal impact, among other mechanisms. A rather large number of methods have been developed to transfer energy for the ionization process, to thermolabile, high- or low-molecular weight, polar or non-polar molecules, in the gas phase (electron impact, EI, chemical ionization, CI, photoionization, PI, field ionization, FI) or in the condensed phase (field desorption, FD, laser desorption, LD, fast atom bombardment, FAB, plasma desorption, PD, secondary ion mass spectrometry, SIMS, matrix-assisted laser desorption ionization, MALDI) [36].

The ionization of neutral molecules is followed by fragmentation, or dissociative ionization, whose product ions can be separated with varying degrees of accuracy, depending on the "spectroscopic balance", i.e, analyzer used. Obtaining a mass spectrum is a *sui generis* energy balance which involves energies ranging from electronic excitation of a molecule (e.g., ultraviolet-visible spectroscopy) up to its atomization (absorption or atomic emission). The ineludible first step is to ionize the molecule (removing or adding an electron, or a proton), which informs about its molecular mass or exact elemental composition when the resulting molecular ion is detected. On the other hand, it is necessary that some of the ionized molecules dissociate or fragment, which reveals what constituent groups comprise the molecule and how they are combined. Electron ionization (EI) is the oldest technique for organic molecule ionization; it is the most widespread and the one with the largest number of applications, for example, GC-MS. It is not so accurate to call it "electron impact", since the size and energy of an electron, compared to an organic molecule of 150-500 Da, does not really allow "impacting" it, such as a ping-pong ball little impact would cause to a rhino. The interaction takes place between the bombarding electron and an electron belonging to the molecule [34-37, 40].

The interaction of an accelerated electron with a molecule leads to the excitation of molecular electrons and a molecular ion, $M^{+\bullet}$, will be formed if the imparted energy permits it, that is, if it is enough to remove an electron from the neutral molecule in the vapor phase. The formed ion's molecular mass is "numerically" equal to that of the molecule because in comparison, an electron mass is negligible. Electrons are excellent agents for ionizing organic molecules. First, because it is easy to obtain them: just pass an electric current through a tungsten or rhenium wire (filament or cathode). Second, their energy is adjustable with the voltage applied between cathode (thermo-electron emitter) and anode (ground connection), where the average standard energy -worldwide accepted convention- is 70 eV. For most organic molecules, the maximum ionization efficiency is already reached with bombarding electrons with energy of 50-60 eV. Standard mass spectra are taken at 70 eV because they so achieve the highest repeatability and reproducibility. Mass spectra libraries are also formed with the spectra obtained by electron ionization energy of 70 eV. All this facilitates the comparison of the spectra taken on different spectrometers with those of databases and other instruments. The bombarding electron energy of 70 eV far exceeds that required to ionize organic molecules. The ionization energies (or ionization potential, IP) for organic molecules lie in the range of 6 to 13 eV, and depend on their molecular structure. For example, to ionize the cyclohexane molecule, 9.9.eV of energy are required; for benzene 9.2 eV, toluene 8.8 eV, pyridine 8.2 eV, and naphthalene 8.1 eV [41,42].

The ionization energy of an organic molecule decreases with the presence therein of π electrons (unsaturated bonds, aromatic ring) or heteroatoms (N, O, S) and with this the ionization cross section of the molecule increases. Small molecules such as H_2O, HCN, CO or CO_2 have quite high ionization energies, namely, 12.6, 13.9, 14.0, and 13.3 eV, respectively (and also their heats of formation are large). When organic molecules are ionized with the loss of 18, 27, 28 or 44 mass units, these small molecular species are practically never recorded as ions in their mass spectra, precisely by having ionization potentials higher than those of any complementary fragment ions, e.g., $(M - H_2O)^{+\bullet}$, $(M - HCN)^{+\bullet}$, $(M - CO)^{+\bullet}$ or $(M - CO_2)^{+\bullet}$. This is also a reflection of the Stevenson-Audier rule [43,44], which states that for a fragmentation process (usually referred to simple bond breaking, but also applied to some simple rearrangement processes) the positive charge is localized on the fragment ion that possesses lower ionization potential. If the molecular ion is not recorded in the mass spectra of volatile analytes and thermally stable substances (alcohols, esters, aliphatic, branched hydrocarbons, etc.), this can be explained by its high lability and very short lifetime ($<10^{-6}$ s), and by the absence in the molecule of structural elements (double bonds, aromatic rings, heteroatoms) to enable stabilization by different mechanisms, generally, through effective charge delocalization. The molecular ion will not become distinguishable in the spectrum by reducing the bombarding electrons energy. If they are not recorded at 70 eV electron energy, they are not detected either on the so-called low voltage spectra, at lower electron energies (10 - 30 eV). This is because the sensitivity of the method falls dramatically, that is, the number of molecular ions or fragment ions decreases with the electron energy. For volatile, thermostable but labile molecules in whose mass spectra no molecular ions are recorded, "soft" ionization methods should be used, e.g., chemical ionization (CI) or in some cases, chemical derivatization turns out to be a useful solution [41].

The electron ionization occurs under reduced pressure (vacuum) of 10^{-5}-10^{-6} Torr (the mean free path is of the order of meters). It is an endothermic process accompanied by the formation of $M^{+\bullet}$ ions and their monomolecular dissociation. However, fragmentation is not a random process. The occurrence of certain fragment ions (fragmentation pattern) is a reflection of: (1) molecular structure, (2) the energy of bombarding electrons, and (3) the excess of internal energy acquired by the ionized molecule. The number of ions (ionic current) detected may be affected by pressure in the system, contamination of the ion source or by aging or deterioration of the dynode surfaces of the electron multiplier. The range of ionization methods for high molecular weight substances (e.g., proteins, nucleotides, polysaccharides, etc.), highly polar molecules (e.g., amino acids, sugars, glycosides, vitamins, pesticides etc.) or thermo-labile substances [40-42] is much larger and technically more sophisticated. Ionization of heavy or polar molecules, usually happens in parallel or immediately after desorption from the surface, where they are bidimensionally distributed as a "solution" in a very low volatile solvent.

Modifiers, polarization promoters, are added or are present as incrustations in a solid [45]. The desorption from the surface and molecular ionization occur almost simultaneously: various primary and secondary ions are generated by means of accelerated atoms (FAB), accelerated ions (SIMS), or high-energy photons (one of the versions is MALDI), which impinge on the sample holder which houses the substance in a properly prepared matrix [46].

Special methods of ionization techniques are used in tandem setups, i.e., in the coupling of liquid chromatography (LC) with mass spectrometry, LC- MS, where the liquid sample could be nebulized in one of the different interfaces available, e.g., assisted by the effect of temperature and high electrical potential; for example, thermospray (TSI) [47], electrospray (ESI) [48], or atmospheric pressure chemical ionization (APCI) [49] methods. Different mechanisms lead to the formation of ionized species, which include molecules with multiple protonation, or deprotonation, clusters, among other ions generated.

The most common technique for the ionization of "small" molecules (more than 90% of applications) is the electron impact, while the positive ion (PICI) or negative ion (NICI) chemical ionization is an important complement used when molecular ions are not recorded in the spectra obtained by EI [50]. When the molecular weight of the substance should be determined, it is clearly deduced from the CI spectra based on the mass of the protonated molecular ion, MH^+ [or deprotonated $(M\text{-}H)^-$, for NICI], along with the cluster ions, usually formed by electrophilic addition of secondary ions from reactant gases (methane, ammonia, *iso*-butane, etc.).

Both electron ionization and chemical ionization are used in the technique of gas chromatography-mass spectrometry. Unfortunately, the use of CI requires, generally, the change of the ionic volume (ionization chamber), because the residual pressures that are used in CI are much higher (up to 1 mm Hg) than those used in EI (10^{-5}-10^{-7} Torr). In fact, less than 10% of all substances in the planet can be ionized in the vapor phase by EI or CI, as the ionization process by these techniques has as major limitation the low volatility or thermal instability of many organic substances. Volatilization is a stage prior to the ionization by EI or CI; the two processes are separate both in time and space. The separation of ions in the GC-MS technique can occur with virtually all types of mass analyzers, e.g., ion traps (IT) [51], quadrupole (Q) [52], magnetic deflection analyzer [53], as well as the configurations of several tandem analyzers (e.g., QqQ, triple quadrupole [54]), and today, for the GC x GC configuration, the time-of-flight (TOF) analyzer [55].

3.2. Mass analyzers

The mass selective detectors are divided into two groups. The first group corresponds to scanning analyzers. These include sector analyzers, e.g., magnetic deflection of single or double focus (when an electrostatic analyzer is added). The magnetic sector analyzer was historically the first mass analyzer employed. However, its use in GC- MS is not common. Quadrupole analyzers are the most frequently used in tandem GC-MS systems. The second group consists of simultaneous ion transmission analyzers. These include the time-of-flight (TOF), different types of ion trap (IT) and Fourier transform mass analyzers, specifically, ion-cyclotron resonance mass spectrometers (ICR-MS), which have gained recent popularity in the field of coupled techniques with both gas and liquid chromatography, because of its high resolution and sensitivity. The mass selective detectors are distinguished by their properties or specifications and analytical scope. The most important parameters include: (1) resolution, (2) the maximum mass that can be measured, and (3) the ion transmission. Low resolution mass analyzers (single focus magnetic deflection, quadrupoles, ion traps, linear TOF) measure nominal masses (integers) and cannot distinguish isobaric species, whereas high resolution analyzers are capable of determining exact masses (4-6 digits after the comma), a measurement that leads to the unambiguous determination of the elemental and isotopic composition. One of the most simple, classic examples, is the measurement of isobaric ions of the species CO^+, N_2^+ and $C_2H_4^+$, whose "separation" is impossible with a low resolution analyzer. It is achieved with high resolution analyzers (magnetic and electric sectors, high-resolution TOF or ICR-MS) working in the exact mass scale [55]. Of course, the gases CO, N_2 and C_2H_2 in a mixture can also be separated and quantified by GC using PLOT type columns. The mass range in which each analyzer operates is important and above all, the maximum mass that it can measure. For example, TOF analyzers have no limits to measure high mass (while the quadrupole and ion trap analyzers do); the mass range of the TOF with reflecton is virtually infinite. The mass range, resolution and accuracy that each analyzer can reach depend on its configuration, the ion separation mechanism and the degree of "monoenergizing" of ions with the same mass, i.e., the ability to decrease the dispersion in time or space, which causes signal (partial ion current) broadening. Ion transmission in mass analyzers is measured as the ratio between the ions formed in the ionization chamber and those which, after passing through the mass analyzer, reach the detector. Scanning analyzers (sector, quadrupole) have lower ion transmission values than the simultaneous transmission analyzers, i.e., TOF, orbitrap, FT-ICR-MS. The latter generally have a higher sensitivity [56].

3.3. Tandem mass spectrometry

Frequently, excessive chemical noise is observed in the ion current of extracts obtained from biological samples, food, soil, etc. This leads to a failure to achieve the required specificity and to detect and identify reliably analytes of interest, when analyzing a complex mixture with many interferences or impurities. Usually, the signal/noise (S/N) increases with the number of steps in an analytical procedure. In instrumental analysis, this is the typical case of a tandem system, for example, GC-MS or LC-MS, including multidimensional or tandem mass spectrometry (MS-MS). Equal to the cleaning of an extract, which increases the S/N in the process of final instrumental analysis, a tandem mass spectrometer includes filtering steps in its operation. For instance, in a triple quadrupole (QqQ), during the first step of ion separation, the first analyzer (MS1) executes a specific clean-up, to distinguish, in a complex mixture, the characteristic ions of the analyte; the second mass analyzer (MS2), records only signals that are characteristic of the target analyte, free from interfering signals. A device located between the two analyzers is where the selected ion can be activated, i.e. its internal energy can be increased, leading to its dissociation and the formation of fragment-ions (ion-products), which are recorded in the analyzer MS2. This device operates as a collisions cell, which cause the disassociation of stable ions selected by the first analyzer.

The classic configuration of a tandem mass spectrometer is the connection in series of MS1, collision- activated chamber and MS2, followed by a system for the detection and measurement of ionic currents. Tandem mass systems are divided into two large groups [38] depending on the types of mass analyzers involved. The first group is made up of tandem-in-time mass spectrometers. These include linear and quadrupolar ion traps, orbital traps (orbitrap) and FT-ICR-MS. In tandem-in-time instruments the ions produced in the ionization region are trapped, are isolated, fragmented and then separated according to their *m/z* ratio in the same physical space. The cascade of dissociation reactions of ions pre-selected and then activated and subsequently monitored, take place in the same analyzer, but occur consecutively as a function of time, thus allowing the successive record of ions which are son, grandson, grand-grandson, etc., $(MS)^n$, of the original selected ion. The second group of tandem mass (MS/MS) instruments is made up of the so-called tandem-in-space mass spectrometers. In these, at least 2 analyzers are separated in space. With these spectrometers it is possible to study not only product-ions, but also precursor ions, the reactions (transitions) between 2 related ions, or monitoring the loss of a neutral fragment. A triple quadrupole, designated as QQQ, or QqQ, belongs to this type of tandem mass spectrometers. Hybrid MS/MS configurations involve combining several analyzers of different nature or different operating principles, for example, quadrupole (Q) or ion trap (IT) with a magnetic sector analyzer (B) alone or in conjunction with electrostatic analyzer (E) or a time-of-flight (TOF) analyzer. This gives rise to different hybrid tandem equipment, e.g., EBE, EBEB, B-QI-Q2, QEB, Q-TOF, IT-TOF-TOFB, EB-TOF-TOF EBE, QBE, and other possible combinations of analyzers. Of course, the union of several analyzers greatly increases the cost of the instrument and the complexity of their operation, but, simultaneously, increases the amount and quality of analytical information obtained, the degree of reliability and specificity [55,56].

Tandem MS/MS instruments involve 2 stages of mass analysis separated by a reaction of activated or induced dissociation of ions. This happens between the mass measurement before and after the fragmentation of the ions selected in the first stage. The fragmentations are caused by collisions of the selected ions with inert gas molecules (He, Ar, Xe or N_2, at a pressure of 0.1-0.3 Pa), and by the accelerating potential of an electrostatic field applied in a collisions cell, also called cell of activated or induced collisions. When employing soft ionization methods (e.g., chemical ionization, Cl) or in the coupling of liquid chromatography with mass spectrometry with electrospray (ESI) or atmospheric pressure chemical ionization (APCI) molecular (or poliprotonated, multicharged) ions practically do not fragment, and this leads to a lack of information required for the structural elucidation of the molecule. The "energization" of the stable (non-dissociated) ions (cations), for example, molecular or quasi- molecular ion, multiprotonated species, cluster, etc., allows to forcibly induce their dissociation to subsequently extract structural information complementary to the molecular mass or elemental composition (when using high resolution mass analyzers).

Using a detection system with more than one mass analyzer such as a triple quadrupole is appropriate for the analysis of target compounds at trace level (ppt - ppb range) in complex matrices, with the presence of interferences, as in the cases of food samples, biological fluids, animal and plant tissues, soil, wastewater and other environmental samples. The MS/MS technique is very helpful and is a valuable analytical reinforcement, required in situations where (1) there is a high chemical noise in spectra acquired in SIM mode, (2) characteristic ions coelute with isobaric impurities (same nominal mass), (3) the structure of the compound is unknown and requires additional structural information (often, it is necessary to activate molecular, quasi-molecular or protonated ions, or some stable fragment ions, to extract additional structural information through the products in which fragment-ions are dissociated), (4) the fragmentogram obtained in SIM mode requires an additional confirmatory information and finally, (5) higher sensitivity and specificity are required in the analysis (pesticide residues, petroleum biomarkers, anabolic steroids and other doping agents, etc.).

The triple quadrupole configuration provides a range of analytical experiments and modes of ionic current acquisition each of which provides certain specific information. The following describes each of the possible modes of acquisition of a QQQ instrument.

3.3.1. Full scan

This is the traditional type of ionic current acquisition, similar to that employed in spectrometers with a single quadrupole analyzer (mass filter). The first analyzer (MS1) performs a full sweep and records mass spectra of every analyte emerging from the GC or LC column, which has been ionized and fragmented into molecular ion and different ion-products. In the GC-MS technique, *ca.* 0.05-1 ng of a compound are sufficient to obtain a mass spectrum that meets quality criteria [34,41]. The intermediate quadrupole (Q2) and the mass analyzer (MS2) operate only in ion transmission mode [52,54].

3.3.2. Selected Ion Monitoring, SIM

This acquisition mode is also well known and is widely practiced in mass spectrometers with a single quadrupole. In this case, the first analyzer (MS1) allows free passage only to a small number of selected ions (usually 3), typical or characteristic of the target analyte, which is selectively sought in a complex mixture. The other two quadrupoles just transmit the ions filtered by MS1. The mass fragmentogram is built based on the partial ion currents recorded. Since each selected ion is measured for a longer period, e.g., approximately 50 ms instead of 50 μs, chemical noise is reduced (S/N is increased), lower detection levels (by a factor of 10-100) are reached, and this permits the detection of a target analyte in quantities of pg or less. SIM mode is widely used for recording both a compound of interest in a complex mixture, as well as for sensitive quantitation, but also to register groups of homologous compounds, by monitoring their characteristic ions, namely for *n*-paraffins: *m/z* 57, 71, 85, for fatty acid methyl esters: ion at *m/z* 74; for alkylbenzenes: *m/z* 91, 105, and for phthalates: *m/z* 149, 167, among other analytes of interest.

3.3.3. Product ion scan

In this acquisition mode a precursor or parent ion is selected in the first mass analyzer or filter (MS1), operating in SIM mode. The chosen parent ion (m^+) passes MS1 and is directed to the collisions cell. The Q2 quadrupole (q) operates only with applied radiofrequency (RF-mode), which allows transmitting ions from the first analyzer to MS2. The collision gas supplied to the cell (usually, He, Ar or N_2), by means of collisions with the selected ions, provides the additional energy (ion excitation process); an applied potential in Q2 (q) permits to accelerate the ions and convert part of their kinetic energy into additional internal energy (rotational, vibrational and electronic). This potential should be optimized in order to obtain higher sensitivity (Figure 8). The increase of internal energy of the ions, that is, their "energizing" or "activation", leads to their dissociation and the formation of different fragment-ions (product ions) which are then directed to the second mass analyzer or filter (MS2). This analyzer scans the mass range smaller than the mass of the selected m^+ parent ion, since the product ions obviously weigh less than their predecessors. It should be noted, that a product-ion mass spectrum will lack the companion signals from isotopes. This acquisition mode is perhaps the most common and well known in the triple quadrupole system, and is used for various purposes and applications. It can be used in the analysis of complex mixtures or substances with impurities, without prior separation in a chromatographic column. If a soft ionization method is used, for example, positive ion chemical ionization, the mixture of ionized substances in the ion source, will produce protonated molecular ions (MH^+). The first mass analyzer MS1 selects one by one the protonated molecular ions, which are directed to the activated collision cell (q), where they dissociate; fragment ions formed from each MH^+ are separated by the MS2 analyzer (Q3) and the spectrum is recorded. In this case, the first analyzer (Q1) acts as a "chromatographic column", the second quadrupole (q) operates as an "ionization chamber", and the MS2 analyzer (Q3) fulfills the role of the mass analyzer, which makes a complete ion scan. Thus, time savings will be achieved by not using a chromatographic column and focusing the QqQ system work in the search of selected compounds of interest present in a complex mixture without prior chromatographic separation.

3.3.4. Parent ion scan

This mode of operation is used to find in a mass spectrum those (precursor) ions that can generate a given fragment (product ion). In this case, the first quadrupole (MS1) operates in the full scan mode, while the third quadrupole (MS2) operates in selected ion monitoring mode. This selected ion is the product ion (daughter ion) of interest, whose precursors are sought. Technically, this takes place as follows: in the MS2 analyzer (Q3) - only product ions with specific mass are filtered, while the first analyzer, MS1, allows passage to all ions with a mass above that of the selected product ion. These ions cross the activated collision chamber (q) where they are fragmented, resulting in, among others, the product ion of interest. MS2 operating in SIM mode is the next filter, which only daughter ions are allowed to cross. For example, in a complex mixture of 30-40 components it is of interest to detect only phthalates (plasticizer). The common ion, diagnostic for this group of compounds, is recorded at *m/z* 149. Then, all ions with the mass/charge ratio higher than *m/z* 149 pass by the first analyzer, operated in scan mode; the analyzer MS2 operates in SIM mode, filtering only ions with *m/z* 149, which are formed in the activated collision chamber from the precursor ions - phthalates. The mass fragmentogram finally recorded contains only chromatographic peaks corresponding to phthalates; thus the mass analyzer, QqQ operated in a precursor ion scanning becomes a selective chromatographic analyzer.

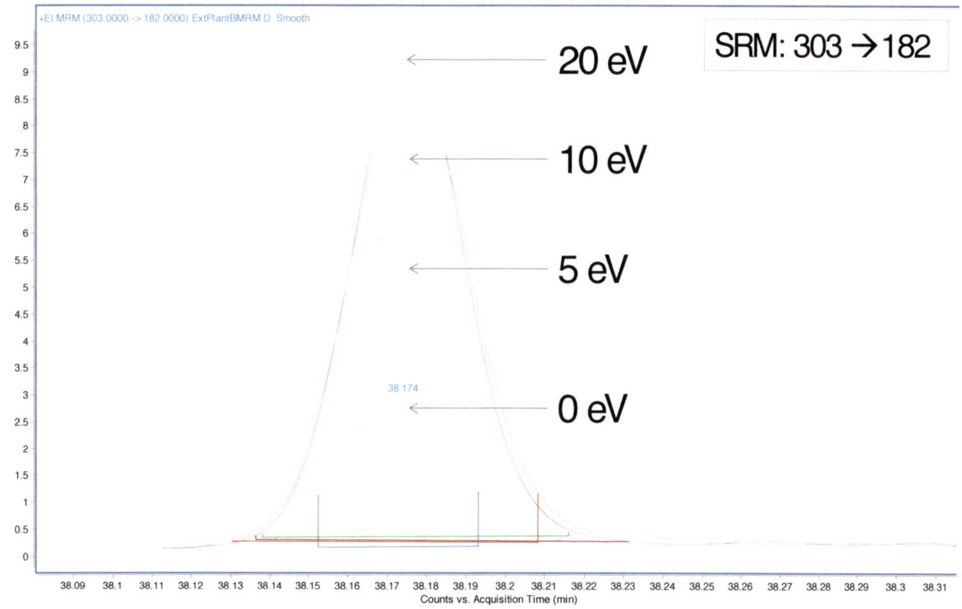

Figure 8. Single reaction monitoring (SRM) (metastable ion transition) between the molecular ion of cocaine (*m/z* 303) and one of the most abundant and characteristic ion-product (*m/z* 182), 303->182, obtained by QqQ-MS with N_2 as collision gas, and at different collision energies (potentials, 0, 5, 10, and 20 eV).

3.3.5. Constant neutral loss scan

In this QqQ operational mode, both mass analyzers, MS1 and MS2, operate in scan mode simultaneously. The scanned masses in both analyzers correspond to ions with a fixed difference, which equals the mass of the selected neutral fragment. If ions f1, f2, f3, f4, etc., pass the first analyzer (MS1), and then go through the activated collision chamber (Q2), they may experience fragmentation. The fragmentation products that will be able to pass the next analyzer (MS2) will be only those ions with the pre-established, fixed, mass difference, that is, f1-Δm, f2-Δm, f3-Δm, f4-Δm, etc. For example, if the linked scan of both analyzers corresponds to a mass difference of 28 units, if ions with *m/z* 80, 81, and 82 pass the first analyzer, MS2 will only allow free passage to ions with *m/z* 52, 53 and 54.

3.3.6. Multiple reaction monitoring, MRM

This is one of the most *sui generis*, interesting, methods of ionic current acquisition of the triple quadrupole, because it allows exploring and properly using its advantages, to turn it into a specific, highly selective and sensitive GC (or LC) mass detector. Unfortunately, when the acquisition of the ionic current is done in the SIM mode, the probability that the selected ion and a signal from the background (chemical noise) actually match is not null. This non negligible possibility of false positive or false negative results, lowers the reliability of the SIM acquisition method. In order to avoid these problems, instead of monitoring characteristic ions, the MRM experiment rather focuses on recording transitions (or transition reactions) between ion pairs (precursor and product). Selected precursor ions (F1) are filtered in the first analyzer (MS1, SIM mode), while only ions F2, product of the transition or dissociation reaction F1 → F2 are allowed to pass the second analyzer MS2 (operating in SIM mode). Both ions must be stable and, in general, abundant in the mass spectrum of the analyte. Reaction monitoring of precursor and daughter (ion product) ions almost completely cancels the probability of coincidence of the analyte signal with one from the background and raises the value of S/N. Typically, the registration of two independent transitions along with chromatographic retention can confirm unequivocally the occurrence of an target analyte in a complex mixture (Figure 9). The use of triple quadrupole operated in MRM mode, is of particular importance for the analysis of compounds (target analytes) present at trace level in highly polluted, complex, multi interference matrices; for example, in pesticide residue analysis in foods, plants or biological or environmental samples [57]. Another important application is the analysis of biomarkers in petroleum [58].

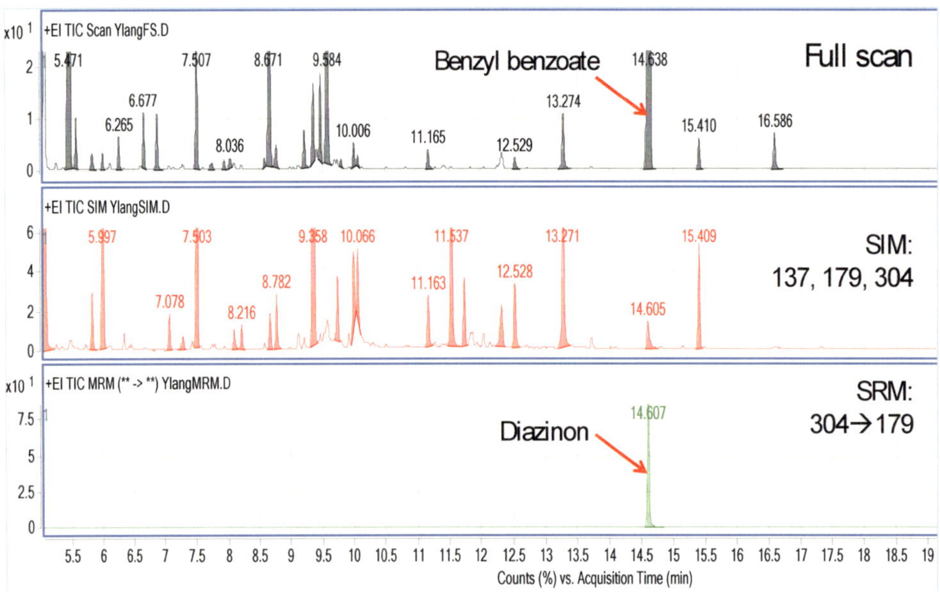

Figure 9. GC-QqQ-MS analysis of the extract obtained from ylang-ylang flowers (*Cananga odorata,* Annonaceae family). Determination of diazinon present in the extract, using different acquisition modes: full scan, SIM [characteristic ions, *m/z* 304 ($M^{+\cdot}$), 179 and 136] and single reaction monitoring, SRM, of the ion transition *m/z* 304 →179. Notice the highly selective and sensitive detection of diazinon (2 ppb) in the ylang-ylang extract, using QqQ-MS in the metastable ion transition mode (SRM).

4. Data analysis and interpretation

There are two basic strategies in GC- MS, for the identification of compounds. The first is the use of standard substances (certified reference material). However, not always all standards are available, many of them are not easily accessible for a large number of analytes. The second strategy is the combination of several approaches, among which are the following: (a) retention indices (RI), in conjunction with (b) experimental mass spectra (EI, 70 eV) and (c) their comparison with those of databases of retention indices obtained in columns of orthogonal polarity (polar and nonpolar) and of standard mass spectra (EI, 70 eV). The combination of several experimental parameters and data, i.e., retention times measured in both columns and mass spectra is mandatory for the structural identification of components in a mixture. The identification can be tentative (preliminary, presumptive) or confirmatory. Confirmation (positive or unambiguous) requires, in many cases, the use of a certified standard compound. Multiple analytes from complex mixtures, however, can have similar retention times or their mass spectra seem alike or have only very small quantitative differences (ion intensities). Limonene epoxides, xylenes, and many structurally similar terpenes, are examples of this situation. However, the possibility of simultaneous coincidence of both the retention indices calculated in both columns (polar and nonpolar) and of mass spectra for two different substances, in fact, is very remote, almost unlikely. For some cases, such as those that may have legal implications, i.e., environmental, forensic cases or disputes, control of doping agents in sports competitions, it is absolutely mandatory to use certified standard substances for identification and confirmation. The analysis of an essential oil, perfume, aroma, and fragrance fractions (or any other complex mixture), in order to quantify and identify its components, done by one-dimensional chromatography, should comply with the following conditions: (1) using preferably long capillary columns (50, 60 m), (2) performing the analysis in two capillary columns with orthogonal phase (e.g., DB-1 or DB-5 and DB-WAX), (3) obtaining experimental mass spectra EI (70 eV) and doing a comparative search, preferably, on various mass spectra databases (e.g., NIST, Wiley, Adams), (4) calculating linear retention indices in two columns, polar and nonpolar, and (5) using standard compounds for further structural confirmation (Figure 10, Table 1). The combination of all these parameters allows confirmatory identification of the mixture components. In GC- MS analysis it is very important to ensure that the chromatographic peaks are "homogeneous", as the co-elution of various substances can lead to structural misassignments.

Compound	Capillary column		
	DB-1	DB-5	DB-WAX
α-Pinene	939-942	932	1036-1038
β-Phellandrene	1025-1032	1025	1213-1216
Sabinene	972-976	969	1130

Table 1. Linear retention indices (LRI) of some monoterpenes, measured on different stationary phase columns.

4.1. Mass spectra interpretation

The large number of fragment ions (cations and cation-radicals) in the mass spectrum is due, firstly, to their formation from molecular ions which have very different excesses of internal

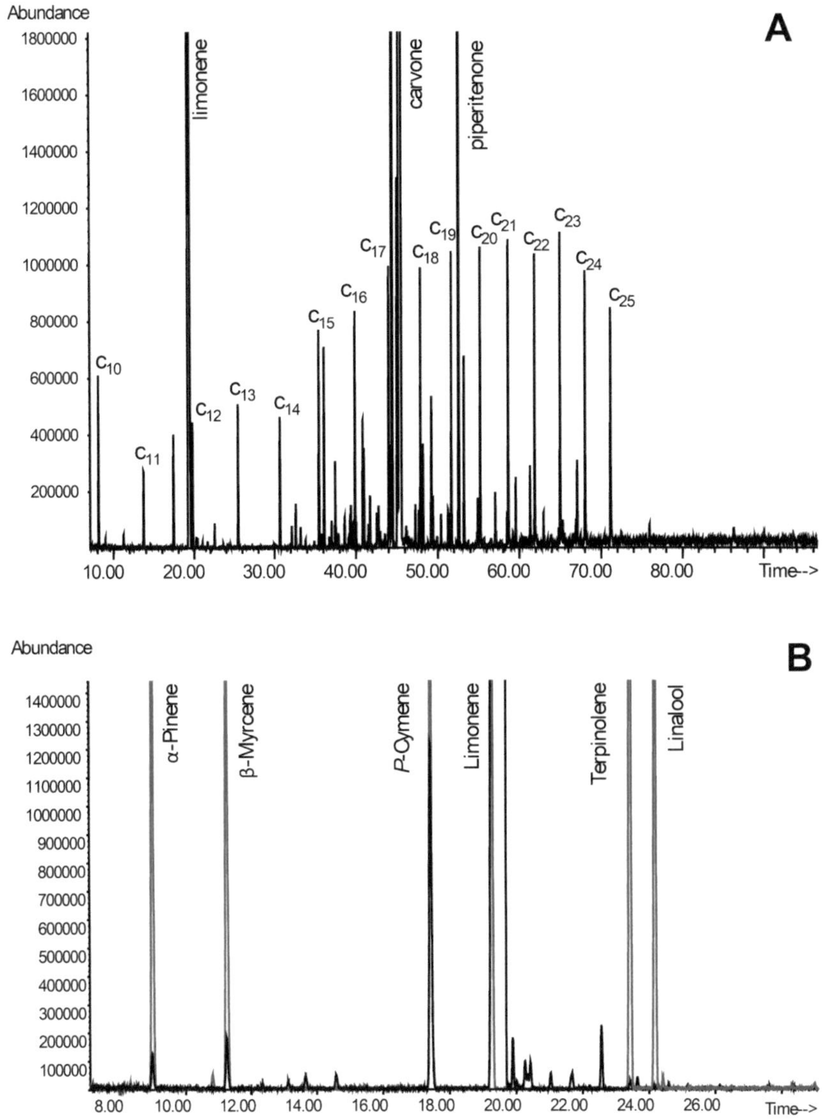

Figure 10. A. The use of *n*-hydrocarbon mixture co-injected with the sample for linear retention indices (LRI) calculation. B. Standard compounds used in the process of sample components identification.

energy. All molecular ions with internal energies lower than the potential energy of appearance of an ion fragment with the lowest formation activation energy will be recorded in the mass spectrum as not dissociated molecular ions. Their intensity in the spectrum depends on the molecular structure and, particularly, their ability to delocalize (stabilize) the positive charge, which allows the ion $M^{+\bullet}$ to exist for longer time than is required for its detection (ca. > 10^{-5} s) in a mass spectrometer. One of the major limitations of the electron ionization technique lies in the fact that molecules that are ionized, should be in the vapor phase, i.e. being volatilizable

http://dx.doi.org/10.5772/57492

without undergoing thermal decomposition, prior to their ionization. Even so, many of the volatilizable and thermostable molecules do not exhibit molecular ions in their mass spectra, only fragment ions, something that limits obtaining information on molecular weight. In fact, less than 10 % of all existing molecules are suitable for analysis by mass spectrometry by electron ionization. The excluded cases are highly polar species (e.g., salts, amino acids), those of high-molecular weight (e.g., proteins, nucleic acids, polymers), and thermolabile compounds (e.g., sugars). In some cases, chemical derivatization of the molecule permits to increase its volatility and thermal stability and decrease its polarity.

4.2. Ion types

The user of mass spectrometry wonders at the beginning, interpreting mass spectra, why molecular ions with the same structure can generate several very different products and can be fragmented by parallel, competitive routes. The type and ratio (abundance) in a mass spectrum of daughter, parent and metastable ions is based on the following factors: (1) distribution of internal energies of molecular ions formed (also valid for fragment ions with excess energy that also dissociate afterwards); (2) activation energy of the process of fragmentation, Ea, and (3) slope of the curve of log K, where K is the dissociation constant of the parent ion through one of the possible reaction coordinates. Rearrangement (or transposition) processes are often accompanied by the formation of intense metastable ions; rearrangement products often "survive" in low voltage (10-15 eV) mass spectra, as the Ea energies for their formation, generally, are lower than those required for a simple bond rupture. As the log K curve becomes steeper, fewer metastable ions are observed in the mass spectrum corresponding to this fragmentation reaction coordinate, and particularly, most likely, this process will be of simple rupture. Accordingly, the structure of the molecule, the excess energy that it acquired during its ionization, the activation energies of possible dissociation processes and their rate, and the frequency factors involved in bond cleavage, are variables that determine the fragmentation pattern of a whole molecule and its product-ions that will be recorded in its mass spectrum. In a mass spectrum, obtained by EI, different signals are recorded, including the molecular ion (if its lifetime is greater than 10^{-6} s), fragment-ions, isotope ions, sometimes, also multicharged ion signals (such as in polyaromatic compounds). Mass spectrometers with specific configurations also detect metastable ions (products of ion fragmentation out of the ionization chamber). Each one of these ions provides structural information more or less important to the structural elucidation of the molecule. Molecular ions indicate the value of molecular mass and the elemental composition when high resolution MS is employed. Intense molecular ion signals indicate that the ionic structure can stabilize its positive charge and this normally suggests aromaticity or highly conjugated unsaturations. It is common that mass spectra of branched hydrocarbons, alcohols, and secondary or tertiary amines are devoid of the molecular ion signal. A soft ionization method, such as chemical ionization, should be used in these cases. Fragment ions contain the primary information for molecular structure elucidation. These may be cations (even electron number) or radical cations (odd electron number). They may result from simple rupture or rearrangement. Certain common fragment losses are associated with the presence of specific groups in the molecule. Alcohol molecular ions easily decay with the loss of either an OH group or an H_2O molecule, giving rise to the

$(M-17)^+$ and $(M-18)^+$ ions. Fragment ions $C_6H_5^+$ and $C_7H_7^+$ dissociate with the loss of an acetylene molecule (C_2H_2) and produce ions at *m/z* 51 and 65, respectively. The substance´s elemental composition is derived from the relative abundance of isotopic ions. The presence of Cl, Br, S or Si can be easily determined from the isotopic peak pattern. A quality requirement of a mass spectrum is the clear distinction of isotopic ions. Multiply charged ions are found in the mass spectra of a limited class of substances, for example, those containing heteroatoms (N, S, O), aromatic or heteroaromatic rings, high unsaturation level, or when several of these elements are combined in the molecule. Their intensity is relatively low and may have fractional values. The signal at *m/z* 64 in the naphthalene mass spectrum (M^+, *m/z* 128) corresponds to a doubly charged molecular ion, M^{2+}. Metastable ions are formed out of the ionization chamber from ions with life time longer than a microsecond, but less than the time required to reach the detector. Their apparent mass m* (a fractional number) relates the masses of parent (m_1) and daughter (m_2) ions according to $m^* = m_2^2/m_1$. Metastable ions are frequently registered in mass spectra obtained with magnetic sector analyzers. A mass spectrum of benzoic acid contains m* signals corresponding to processes of successive loss of radical OH and the CO molecule: $M^+ \rightarrow (M - OH)^+$ and $(M - OH)^+ \rightarrow [(M - OH) - CO]^+$. These signals confirm the genetic link between these ions. The absence of an m* ion corresponding to the direct process $M^+ \rightarrow (M - COOH)^+$ indicates that this process does not take place.

4.3. Fragmentation patterns

The fragmentation pattern (*m/z*, the amount, I%, of ions that are recorded in a mass spectrum) shows how a molecule dissociates once ionized; this does not occur randomly, because it is derived from the molecular structure (nature of bonded atoms, chemical bonding strength), their spatial arrangement, ionization potential and internal energy that acquires during its collision with an electron. The fragmentation pattern is unique to each molecular structure and contains clearly detectable differences (often only quantitative), including those for the isomers (positional, geometric or stereoisomers). The mass spectrum provides information on: (1) molecular mass: if the molecular ion does not appear in the EI spectrum, it is necessary to obtain the spectrum via soft ionization, e.g., CI, but sometimes, derivatization of the molecule [35,37] can be used, in order to find its molecular weight and also to increase its volatility (that of the derivative) and the sensitivity with which it can be detected; (2) elemental composition: it is possible to know it, based on the analysis of a mass spectrum obtained by a high resolution mass spectrometer. Isotope ions or typical fragment losses (-HCN,-CO,-OH, -HS,-NH_2, etc.) also allow inferences about the elemental composition of the substance; the simplest case is the presence of Cl or Br atoms in the molecule; (3) functional groups in the molecule can be elucidated by noticing in the spectrum signals corresponding ions with characteristic mass, which indicate their presence, e.g., phenyl (*m/z* 77), benzyl (*m/z* 91), benzoyl, C_6H_5CO, (*m/z* 105), or by detecting fragments, the products of the loss of small neutral molecules, e.g., water, CO_2, CO, C_2H_2, HCN, alkyl radicals (CH_3, C_2H_5, C_3H_7, etc.); these fragment ions are related to the presence of these structural groups in the molecule; (4) spatial structure: sometimes it is possible to establish the spatial configuration based on the mass spectrum, especially when the spectra of both stereoisomers can be compared or when using low voltage mass spectra (10-20 eV) or chemical ionization, on which the differences in conformational energy (enthalpy

http://dx.doi.org/10.5772/57492

of formation) can be observed more clearly, based on differences in intensities of characteristic fragment ions, or in ionization potentials. Generally, dissociative ionization processes can be divided into the following two major groups: (1) single rupture reactions (homolytic or heterolytic) and (2) rearrangement reactions, which can be rearrangement of the molecular skeleton (for example, tropylium ion formation, $C_7H_7^+$) or reactions accompanied by hydrogen transpositions, for example, the classic McLafferty rearrangement [34]. The most common single rupture reactions are cation formation (for example, in hydrocarbon chains), allylic and benzylic ruptures, retro-Diels-Alder (RDA) reaction in monounsaturated cyclic systems, and acyl ion formation, among others. Generally, the energy required (activation energy, Ea, of the process) for rearrangement is lower than that required for single rupture. Single rupture processes are generally less selective than the molecular skeleton or hydrogen transpositions, and are also less susceptible to steric effects in the molecule compared to rearrangement processes. The mass spectra obtained with 70 eV or lower voltage (10-30 eV) of the same substance differ markedly. In the latter, molecular ions prevail (relative increment) and some products survive, basically, hydrogen rearrangements or transpositions, which require lower activation energy. The binding energy of atoms in the ionized molecule is in the order of 2-4 eV, whereas the ionization energy of the organic molecule is of the order of 6-13 eV. The energy of the bond involved in the dissociation is an important factor, especially when there are competitive processes to form predominantly one product versus another. The interpretation of mass spectra of organic molecules, obtained by electron ionization, is based on the study of the fragmentation pattern, which depends on molecular ion and ion-fragment dissociation mechanisms. Fragmentation schemes are not necessarily speculative, because numerous experimental verifications of fragmentation routes can be found. Different methods are included within the experimental tools to establish fragmentation schemes and mechanisms: (1) using soft (CI, FI) ionization techniques, (2) chemical derivatization, (3) isotopic labeling (e.g., with D, ^{15}N, ^{18}O) and chemical marking (introduction to the molecule of substituents that are not released during the fragmentation, e.g., F, OCH_3, etc.); (4) high resolution mass spectrometry, which allows to establish the elemental composition of each ion; (5) metastable ion study to determine the relationship between parent ions and subsidiary ions, (6) use of tandem configurations and study of collision-activated dissociation (CID, Collision-Induced Dissociation); (7) photoionization (PI) and determination of energies (potential) of ionization and ion appearance; (8) quantum mechanical calculations of formation energies and ion structures. Obviously, not all experimental techniques can be performed on a single mass spectrometer, but the whole set will provide sufficient information to correctly set the scheme or even the mechanism for the fragmentation of an organic molecule ionized by electrons. Many references [34,35,36,41,42,48,52], contain information about fragmentation of organic molecules, their mechanisms and the classification of simple rupture or transposition processes. Generally, it is considered that the energies involved during the electron ionization are very high (70 eV) and can "level" or mask thermodynamic differences generated by the spatial characteristics in isomeric molecules. However, steric factors exert pronounced effects on activated complex structure and are more relevant to the rearrangement processes. Increasing the internal energy of ion $M^{+\bullet}$ causes a decrease of rearrangement products. Decreasing the energy of bombarding electrons to 10-20 eV, can reduce the internal energy of the ions $M^{+\bullet}$ and increase the number of rearrangement products (transpositions). Stereoisomeric effects are demonstrated most notably in the mass spectra of chemical ionization or those taken in tandem configuration instruments when collision activated reactions (CID) are used.

5. Conclusion

The association of separation and mass spectrometric techniques is a key that opens up a rich and multidimensional analytical space for the investigation of complex mixtures with high sensitivity, selectivity and specificity. The variation of chromatographic conditions and the modulation of mass spectrometric data acquisition or the use of tandem spectrometers, permit to focus the analytical methods towards different goals in these investigations. Trace-level contaminant detection, molecular structure characterization, or classification, are just a few of the many applications of gas chromatography coupled to mass spectrometry performed currently with varying degrees of resolution and sensitivity on very complex mixtures.

Author details

Elena Stashenko* and Jairo René Martínez

*Address all correspondence to: elena@tucan.uis.edu.co

Research Center for Biomolecules, CIBIMOL, CENIVAM, Universidad Industrial de Santander, Bucaramanga, Colombia

References

[1] Chaintreau A. Simultaneous Distillation–Extraction: From Birth to Maturity - Review Flavour and Fragrance Journal 2001; 16(2) 136-148.

[2] Flotron V, Houessou JK, Bosio A, Delteil C, Bermond A, Camel V. Rapid Determination of Polycyclic Aromatic Hydrocarbons in Sewage Sludges Using Microwave-Assisted Solvent Extraction. Comparison with Other Extraction Methods. Journal of Chromatography. A 2003; 999(1-2) 175-84.

[3] Kingston HM, Haswell SJ. Microwave-Enhanced Chemistry. Washington:American Chemical Society; 1997.

[4] Rice SL, Mitra S. Microwave-Assisted Solvent Extraction of Solid Matrices and Subsequent Detection of Pharmaceuticals and Personal Care Products (Ppcps) Using Gas Chromatography–Mass Spectrometry. Analitica Chimica Acta 2007; 589 125-132.

[5] Cabaleiro N, de la Calle I, Bendicho C, Lavilla I. Current Trends in Liquid-Liquid and Solid-Liquid Extraction for Cosmetic Analysis: A Review. Analytical Methods 2013; 5(2) 323-340.

[6] Fritz JS. Analytical Solid-Phase Extraction. New York: Wiley-VCH; 1999.

[7] Sairam P, Ghosh S, Jena S, Rao KNV, Banji D. Supercritical Fluid Extraction (SFE)-An Overview. Asian Journal of Research in Pharmaceutical Science 2012; 2(3) 112-120.

[8] Ge X, Lv Y, Wang A. Application of Accelerated Solvent Extraction in the Analysis of Organic Contaminants, Bioactive and Nutritional Compounds in Food and Feed. Journal of Chromatography A 2012; 1237 1–23.

[9] Risticevic S, Vuckovic D, Pawliszyn J. Solid-Phase Microextraction. In: Pawliszyn J, Lord H. (eds.) Handbook of Sample Preparation. Hoboken, NJ, USA: John Wiley & Sons, Inc. 2010. doi: 10.1002/9780813823621.ch5.

[10] Heaven MW, Nash D. Recent Analyses Using Solid Phase Microextraction in Industries Related to Food Made into or From Liquids. Food Control 2012; 27(1) 214–227.

[11] Stashenko E, Martínez J. Derivatization and Solid-Phase Microextraction. Trends in Analytical Chemistry 2004; 23(4) 553-561.

[12] Tholl D, Boland W, Hansel A, Loreto F, Rose USR, Schnitzler JP. Practical Approaches to Plant Volatile Analysis. The Plant Journal 2006; 45(4) 540-560.

[13] Kolb B, Ettre L. Static Headspace-Gas Chromatography. New York: Wiley-VCH; 1997.

[14] Narendra N, Galvão MDS, Madruga MS. Volatile compounds Captured Through Purge and Trap Technique in Caja-umbu (*Spondias* sp.) Fruits During Maturation. Food Chemistry 2007; 102 726-731.

[15] Musshoff F, Lachenmeier DW, Kroener L, Madea B. Automated Headspace Solid-Phase Dynamic Extraction for the Determination of Cannabinoids in Hair Samples. Forensic Science International 2003; 133 32-38.

[16] Bagheri H, Babanezhad E, Khalilian F. An Interior Needle Electropolymerized Pyrrole-Based Coating for Headspace Solid-Phase Dynamic Extraction. Analytica Chimica Acta 2009; 634(2) 209-214.

[17] Bicchi C, Cordero C, Liberto E, Rubiolo P, Sgorbini B, David F, Sandra P. Dual-phase twisters: A new approach to headspace sorptive extraction and stir bar sorptive extraction. Journal of Chromatography A 2005; 1094(1-2) 9-16.

[18] C. Cordero, E. Liberto, P. Rubiolo, B. Sgorbini, P. Sandra. Sorptive tape extraction in the analysis of the volatile fraction emitted from biological solid matrices. Journal of Chromatography A 2007; 1148(2) 137–144.

[19] Grob K. Split and Splitless Injection in Capillary GC, 3d Ed, Heidelberg: Huthig Buch Verlag; 1993.

[20] Grob K. On-Column Injection in Capillary Gas Chromatography. 2d Ed. Heidelberg: Dr. Alfred Heuthig, Gmbh; 1987.

[21] Jennings W, Mittlefehldt E, Stremple P. Analytical Gas Chromatography, 2nd Ed, San Diego: Academic Press; 1997.

[22] Hinshaw JV, Ettre LS. Introduction to Open-Tubular Column Gas Chromatography. Cleveland: Advanstar; 1994.

[23] Scott RPW. Introduction to Analytical Gas Chromatography. 2d Ed, New York: Marcel Dekker; 1998.

[24] Rood D. A Practical Guide to the Care, Maintenance, and Troubleshooting of Capillary Gas Chromatographic Systems. Weinheim: Wiley-VCH; 1999.

[25] Adlard ER. Chromatography in the Petroleum Industry. Amsterdam: Elsevier; 1995.

[26] Beesley TE, Scott RP. Chiral Chromatography. Chichester: John Wiley; 1998.

[27] Es AV. High Speed Narrow Bore Capillary Gas Chromatography. Heilderberg: Hüthig; 1992.

[28] Domotorova M, Matisova E. Fast Gas Chromatography for Pesticide Residues Analysis. Journal of Chromatography A 2008; 1207(1-2) 1-16.

[29] Tranchida PQ, Mondello L. Current-Day Employment of the Micro-Bore Open-Tubular Capillary Column in the Gas Chromatography Field. Journal of Chromatography A 2012; 1261 23–36.

[30] Marsman JH, Wildschut J, Evers P, de Koning S, Heeres HJ. Identification and Classification of Components in Flash Pyrolysis Oil and Hydrodeoxygenated Oils by Two-Dimensional Gas Chromatography and Time-of-Flight Mass Spectrometry. Journal of Chromatography A 2008; 1188(1) 17-25.

[31] Stashenko E. Aceites Esenciales. Bucaramanga: Ediciones UIS; 2009.

[32] Mondello L, Tranchida PQ, Dugo P, Dugo G. Comprehensive two-dimensional gas chromatography-mass spectrometry: A review. Mass Spectrometry Reviews 2008; 27(2) 101–124.

[33] Dressler M. Selective Gas Chromatographic Detectors. Amsterdam: Elsevier; 1986.

[34] Hubschmann HJ. Handbook of GC-MS: Fundamentals and Applications. 2d Ed. Weinheim: Wiley-VCH; 2009.

[35] Laskin J, Lifshitz C. Principles of Mass Spectrometry Applied to Biomolecules. New York: John Wiley and Sons; 2006.

[36] Ashcroft A. Ionization Methods in Organic Mass Spectrometry. London: Royal Society of Chemistry; 1997.

[37] Ahuja S, Jespersen ND. Modern Instrumental Analysis. Amsterdam: Elsevier; 2006.

[38] Kandiah M, Urban PL. Advances in Ultrasensitive Mass spectrometry of Organic Molecules. Chemical Society Reviews 2013; 42 5299-5322.

[39] Kataria S, Prashant B, Akanksha M, Premjeet S, Devashish R. Gas Chromatography-Mass Spectrometry: Applications. International Journal of Pharmaceutical & Biological Archives 2011; 2(6) 1544-1560.

[40] McMaster M. GC/MS: A Practical User´s Guide. New York: John Wiley and Sons; 2011.

[41] Dass C. Fundamentals of Contemporary Mass Spectrometry. Hoboken: John Wiley and Sons; 2007.

[42] Sparkman OD, Penton Z, Kitson FG. Gas Chromatography and Mass Spectrometry: A Practical Guide. Oxford: Academic Press; 2011.

[43] Stevenson DP. Ionization and Dissociation by Electronic Impact. The Ionization Potentials and Energies of Formation of Sec.-Propyl and Tert.-Butyl Radicals. Some Limitations on the Method. Discussions of the Faraday Society 1951; 10 35-45.

[44] Audier HE. Ionisation et fragmentation en spectrometrie de masse—I: Sur la répartition de la charge positive entre fragments provenant de mêmes ruptures. Organic Mass Spectrometry 1969; 2(3) 283-298.

[45] Laiko VV, Baldwin MA, Burlingame AL. Atmospheric Pressure Matrix-Assisted Laser Desorption/Ionization Mass Spectrometry. Analytical Chemistry 2000; 72(4) 652–657.

[46] Qian K, Edwards KE, Siskin M, Olmstead WN, Mennito AS, Dechert GJ, Hoosain NE. Desorption and Ionization of Heavy Petroleum Molecules and Measurement of Molecular Weight Distributions. Energy & Fuels, 2007; 21(2) 1042–1047.

[47] Blakley CR, Vestal ML. Thermospray Interface for Liquid Chromatography/Mass Spectrometry. Analytical Chemistry 1983; 55(4) 750–754.

[48] Sleighter RL, Hatcher PG. The Application of Electrospray Ionization Coupled to Ultrahigh Resolution Mass Spectrometry for the Molecular Characterization of Natural Organic Matter. Journal of Mass Spectrometry 2007; 42(5) 559–574.

[49] Van Berkel GJ, Pasilis SP, Ovchinnikova O.Established and Emerging Atmospheric Pressure Surface Sampling/Ionization Techniques for Mass Spectrometry. Journal of Mass Spectrometry 2008; 43(9) 1161–1180.

[50] Meyer MR, Maurer HH. Current Status of Hyphenated Mass Spectrometry in Studies of the Metabolism of Drugs of Abuse, Including Doping Agents. Analytical and Bioanalytical Chemistry 2012; 402 195-208.

[51] March R, Todd JFJ. Practical Aspects of Ion Trap Mass Spectrometry: Ion Trap Instrumentation. Boca Raton: CRC Press; 1995.

[52] Dawson PH. Quadrupole Mass Spectrometry and Its Applications. Heidelberg: Springer; 1995.

[53] Brenton AG, Godfrey AR, Alamri M, Stein BK, Williams CM, Hunter AP, Wyatt MF. Analysis of large historical accurate mass data sets on sector mass spectrometers. Rapid Communications in Mass Spectrometry 2009; 23(21) 3484–3487.

[54] Belov ME, Prasad S, Prior DC, Danielson WF, Weitz K, IbrahimYM, Smith RD. Pulsed Multiple Reaction Monitoring Approach to Enhancing Sensitivity of a Tandem Quadrupole Mass Spectrometer. Analytical Chemistry 2011; 83 2162–2171.

[55] Xian F, Hendrickson CL, Marshall AG. High Resolution Mass Spectrometry. Analytical Chemistry 2012; 84 708–719.

[56] El-Aneed A, Cohen A, Banoub J. Mass Spectrometry, Review of the Basics: Electrospray, MALDI, and Commonly Used Mass Analyzers. Applied Spectroscopy Reviews 2009; 44(3) 210-230.

[57] Wang X, Wang S, Cai Z. The Latest Developments and Applications of Mass Spectrometry in Food-Safety and Quality Analysis. TrAC Trends in Analytical Chemistry 2013; 52 170–185.

[58] Eiserbeck C, Nelson RK, Grice K, Curiale J, Reddy CM. Comparison of GC–MS, GC–MRM-MS, and GC × GC to Characterise Higher Plant Biomarkers in Tertiary Oils and Rock Extracts. Geochimica et Cosmochimica Acta 2012; 87 299–322.

Chapter 2

Analytical Aspects of the Flame Ionization Detection in Comparison with Mass Spectrometry with Emphasis on Fatty Acids and Their Esters

Jesuí Vergilio Visentainer, Thiago Claus,
Oscar Oliveira Santos Jr,
Lucas Ulisses Rovigatti Chiavelli and
Swami Arêa Maruyama

Additional information is available at the end of the chapter

http://dx.doi.org/10.5772/57333

1. Introduction

The initial development of gas chromatography (GC) is deeply interconnected with lipid analysis. This separation technique may be considered as the main contribution for most of knowledgement regarding fatty acid (FA) composition that exists today. Actually, it is possible to convert milligrams of a lipidic sample to fatty acid methyl esters (FAMEs), separate them in a gas chromatograph and quantify these fatty acids in a short time [1].

Probably, the most important identification method which is coupled with GC is mass spectrometry (MS). It is older than GC, and this fact might be surprising for many readers. However, the basic principles and the first separations of atomic masses were primarily demonstrated in the last part of 19th century, while chromatographic columns appeared during 20th century [2].

The possibility of coupling a mass spectrometer in the exit of a chromatographic column, along with the advent of modern computers, introduction of fused silica capillary columns and reduction of interference issues allowed the achievement of analytical results which are more precise/accurate. Allied to these facts, the reduction of gas chromatographs prices enforced their application as a tool for researches with lipidic materials. However, in relation to food analysis with emphasis on fatty acid composition, the flame ionization detector (FID) is the most popular one, because it is easy to operate, possesses a wide linearity range, rapid response, and its limit of detection is of 10^{-12}g for alkanes (an almost universal answer) [3].

In the last years, it can be observed a significant increase of published works about FAs which also envolve new discoveries regarding good and bad effects of FAs for human health, new food products derived from fats and oils available for the end user, novel FAs (such *trans* FAs) which do not exist in certain foods, supplementation of FAs in food products and optimization of biodiesel synthesis.

1.1. Derivatization of functional groups

Before analyzing FAs through GC, it is necessary to convert them into compounds with greater volatility and thermal stability (esters, for example). The transformation of FAs into other organic functionalities with lower boiling points enhances their elution through a chromatographic column. Thus, if correct conditions are used, it is possible to separate compounds with close molar masses or even isomers [3].

The formation of these esters commonly occurs from reactions between lipids with short chain monoalcohols (like (m)ethanol) which provide (m)ethyl radical. The competition between use of methanol versus ethanol originated from factors such as disponibility, reaction yield, toxicity, among others.

A method which is very efficient for fatty acid methylation was proposed in works done by Joseph and Ackman (1992) [4], and it is commonly employed as a fine tool for studies of lipidic composition.

1.2. Integration of chromatographic peaks

After sample injection in a gas chromatograph (GCph) and further separation of analytes through the chromatographic column, an chromatogram of the sample of interest is obtained. Such chromatogram is produced by the integration of the signals which are collected by the detector in a measure related with the amount of each analyte which are detected. In FID case, compounds are burned. Besides, important information such as peak number/area/percentages, as well as their respective retention times, can be obtained from the chromatogram [5].

These integration measures may be related with the existing FAs or FAMEs concentrations in a food/biodiesel sample. However, in order to apply this relation between chromatographic peak area and analyte concentration, the analyst must understand the principles and execute the necessary corrections. Otherwise, the expressed results will be erroneous and will not be clearly understood [6].

1.3. Flame ionization detector (FID) and the differential response

The used of FID for quantification of fatty acid esters (FAEs) is advantageous in relation to other detector types, due to the reasons mentioned before (it is easy to operate, possesses a wide linearity range, rapid response, and its limit of detection is of 10^{-12}g for alkanes). For a didactical approach, from now on the discussions will be centered on FAMEs, although this book chapter may be extended for every FAE (ethyl esters, propyl esters, etc.) [5].

Instead of responding to solute property, FID is sensitive to the mass flux which passes through it in a determined amount of time. It is important to point out that, despite the fact that a solution is injected, solvent and methyl esters are separated by the column during chromatographic run. Thus, a certain mass, in gaseous form, of each compound will arrive at the detector togethter with the carrier gas. Inside FID, it undergoes combustion in a flame produced by oxygen and hydrogen, producing ions which will be collected by electrodes. The amount of formed ions when a certain methyl ester is present in carrier gas is greater than the amount of ions which are formed when only carrier gas is being burned, thus the generated current is converted into a voltage, amplified and registered under the form of a chromatogram [5].

Reaction (1) demonstrates the chemical ionization that occurs in FID:

$$CH + O \rightarrow CHO^{+} + e^{-} \qquad \text{(Reaction 1)}$$

The CHO^{+} ion is unstable and quickly reacts with water in the flame in order to produce hydronium, according with the reaction (2):

$$CHO^{+} + H_2O \rightarrow H_3O^{+} + CO \qquad \text{(Reaction 2)}$$

This reaction occurs for every 100000 carbon atoms which are introduced in the flame [3]. Thus, FID response is proportional to the number of carbon atoms which are burned.

Specifically, the magnitude of the signal generated by FID is proportional to the number of carbon atoms which are bonded to hydrogen atoms (active carbon, C*). However, there are intrinsic factors of a FAME molecule that alters FID response, such as presence of oxygen, which diminishes such response. The carbon atom of carboxylate group (COO) is not ionized in a proper manner during combustion, thus it is not considered as active carbon.

1.4. Mass spectrometric detector

Mass spectrometry is one of the most versatile and sensitive analytical methods which is employed in many studies, ranging from medical to technological science. This method allows the molecular mass/structure determination of unknown compounds, and to quantify small molecules to biomolecules such as proteins, lipids and oligonucleotides [7].

In MS analysis, samples may be directly inserted in the mass spectrometer (off-line analysis) or this equipment may be coupled with a separation technique such as GC, liquid chromatography (LC) or capillary electrophoresis (CE) [7].

A mass spectrometer is basically composed of an ionization source, where the analytes of a sample are ionized and/or fragmented in ions with specific values of mass/charge ratio (m/z). They are immediately accelerated towards a mass analyzer, which has the function of selecting ions according to the desired m/z values. Such selection occurs through the application of electric and/or magnetic fields. A detector transforms the ionic (fragment) signals to electrical signals, and the magnitude of these signals in function of m/z ratio is converted by a data processor, generating a corresponding mass spectrum. The electron multipliers are among the most frequently used detectors in mass spectrometers [8].

The advanced methods in MS differ especially in sample ionization modes and mass analyzers. The most suitable ionization mode for an analysis will depend on physicochemical properties of analytes, such as polarity, molecular mass and boiling point [8], etc.

2. Methods of relative normalization and absolute quantification

Also called area normalization, this method is based on the relative area percentage of a certain FA in relation to the total area of all fatty acids which are eluted from the column. Despite the fact it is an obsolete method, a lot of works regarding FAEs analysis in foods/biodiesel still use it, where the results are expressed in relative area percentage. The disadvantages of normalization are propagation of errors due to results interdependence. Every component of a sample must be detected and, if a certain compound is omitted/estimated, the areas of other substances are affected. Besides, results obtained through this method normally are difficult to interpretand published in an erroneous form.

However, in the expression of results in form of FAMEs (biodiesel) or FAs (foods), where concentrations are expressed in FAME mass *per* raw material/sample mass, using internal standard (IS) or correction factors for FID will lead to more accurate results which can be used by professionals from different areas without major difficulties. It is noteworthy that, if correction factors are complicated for use, internal standard calibration must be used to express results as concentrations, not as area percentages [5].

3. Transformation of methyl ester percentual area in fatty acid concentration: an alternative method

In this method, conversion factors are used in the transformation of relative percentage of a chromatographic peak from a methylic ester (normalization method) in mass amounts of the corresponding FA. Results commonly are expressed in (milli)grams of FA *per* 100g of food or total lipids. Such a conversion factor is obtained from the amount in mass of FAs (tabulated values) which exists in the different classes of total lipids (triacylglycerols and phospholipids, mainly) and form their different masses in a certain food sample.

Some authors describe that the average percentual mass contribution of FAs in triacylglycerols (TAGs) and phospholipids (PLs) corresponds to 95.6% and 72.0%, respectively. This means that, upon expressing values in decimals, it can be understood that 1g of phospholipids possess about 0.720g of FAs. Thus, the determination of F factor (decimal) may be done in the following way [6]:

F (decimal) = TAG × 0.956 + PL × 0.720

Where:

TAG = Mass amount of TAGs in total lipids.

PL = Mass amount of PLs in total lipids.

Assuming that a total lipid sample contains 65% in mass of TAG and 30% in mass of PL, the F factor with application decimal values will be:

F (decimal) = 0.65 × 0.956 × 0.30 × 0.720

= 0.84

Knowing that the mass percentage of a certain X fatty acid (relative percentage by normalization method) in sample is 26%, the amount in grams of this acid *per* 100g of total lipids (TLs) will be:

g of fatty acid X/100 g of TL = F (decimal) × % Area

g of fatty acid X/100 g of TL = 0.84 × 26

g of fatty acid X/100 g of TL = 21.76 g.

The conversion factor may be found for some specific foods. For example, Table 1 shows the F conversion factors (decimal) for different classes of TAGs and PLs in total lipids of fish.

Despite this method being easy and practical, it includes some constant values, often theoretical, that might result in errors. These constants and other interpretations will be discussed later in a practical example.

4. Internal standard calibration and correction factors

In order to obtain greater accuracy from quantification, it must be taken into account that FAEs respond to FID in a differential manner. Thus, it is necessary to use the theoretical correction factor (F_{CT}), determined through the number of active carbons (C*), and the experimental/empiric correction factor (F_{CE}). Both factors are obtained from comparisons between FAMEs and an internal standard [3].

The internal standard calibration method is considered less sensitive to injection errors and instrumental variations. Besides, the internal standard must have high degree of purity/stability, must not be component of sample, and it needs to elute separately from other analytes. It is not easy to fulfill all these requirement during a FAEs mixture analysis. However, some esters have been recommended for use as IS, such as methyl tricosanoate (23:0Me) for foods and methyl heptadecanoate (17:0Me) / ethyl oleate (18:1Et) for biodiesel, because they

% in mass of TL	% TAG	% PL	F (decimal)
0.65	-	92.3	0.66
0.70	7.1	85.7	0.68
0.80	18.8	75.0	0.72
0.90	27.8	66.7	0.75
1.00	35.0	60.0	0.77
1.25	48.0	48.0	0.80
1.50	56.7	40.0	0.83
1.75	62.9	34.3	0.85
2.00	67.5	30.0	0.86
2.50	74.0	24.0	0.88
3.00	78.3	20.0	0.89
3.50	81.4	17.1	0.90
4.00	83.8	15.0	0.91
4.50	85.6	13.3	0.91
5.00	87.0	12.0	0.92

% in mass of TL = percentage in mass of total lipids per 100g of food; %TAG and %PL= percentage in mass of triacylglycerols and phospholipids, respectively; F (decimal)= decimal factor. Source: Kinsella [9] (apud EXLER, 1975, p. 154).

Table 1. Conversion factors based on the total lipid content.

are not naturally found in the samples cited above, possess good stability, and do not contain unsaturation in their carbonic chains[9].

4.1. Theoretical correction factor (F_{CT})

The determination of F_{CT} is based on two facts: (1) that FID proportionally responds to the relative mass percentage of a FAME carbonic chain and (2) this detector does not respond, during combustion, in a proper way to carbon of carboxylate group (COO^-). Thus, randomly assigning a F_{CT}=1 for a fame, for example, methyl stearate (18:0Me), the correction factors of other methyl esters can be calculated, according to Table 2. Thus, FID`s F_{CT} is a constant and it can not be modified due to instrumental errors [3].

4.2. Theoretical correction factor (F_{CT}) determination and calculus of active carbon (C*) percentage

In order to show the application of active carbon (C*) percentage determination to FID`s $F_{CT}$, methyl stearate ($C_{19}H_{38}O_2$, molar mass=298.5080g) was used as internal standard to calculate methyl decanoate`s factor ($C_{11}H_{22}O_2$, molar mass=186.2936g). It is important to remind that the

*Methyl ester	Methyl esters of reference as internal standard						
	18:0 Methyl stearate	**12:0 Methyl dodecanoate**	**17:0 Methyl heptadodecanoate**	**18:1 Methyl oleate**	**19:0 Methyl nonadodecanoate**	**21:0 Methyl heneicosanoate**	**23:0 Methyl tricosanoate**
4:0	1.5396	1.4294	1.5257	1.5501	1.5522	1.5742	1.5930
5:0	1.4009	1.3006	1.3883	1.4105	1.4123	1.4324	1.4495
6:0	1.3084	1.2147	1.2966	1.3174	1.3191	1.3378	1.3538
7:0	1.2423	1.1534	1.2311	1.2508	1.2524	1.2702	1.2854
8:0	1.1927	1.1073	1.1820	1.2009	1.2024	1.2195	1.2340
9:0	1.1542	1.0716	1.1438	1.1621	1.1636	1.1802	1.1942
10:0	1.1233	1.0429	1.1132	1.1310	1.1325	1.1486	1.1622
11:0	1.0981	1.0195	1.0882	1.1056	1.1071	1.1228	1.1361
12:0	1.0771	**1.0000**	1.0674	1.0845	1.0859	1.1013	1.1144
12:1	1.0670	0.9906	1.0574	1.0743	1.0757	1.0910	1.1040
13:0	1.0593	0.9835	1.0497	1.0666	1.0680	1.0831	1.0960
14:0	1.0441	0.9694	1.0347	1.0512	1.0526	1.0676	1.0803
14:1	1.0354	0.9613	1.0261	1.0425	1.0439	1.0587	1.0713
15:0	1.0308	0.9570	1.0215	1.0379	1.0392	1.0540	1.0665
15:1	1.0227	0.9495	1.0135	1.0297	1.0311	1.0457	1.0581
16:0	1.0193	0.9463	1.0101	1.0263	1.0276	1.0422	1.0546
16:1	1.0117	0.9393	1.0026	1.0186	1.0200	1.0345	1.0468
16:2	1.0041	0.9322	0.9951	1.0110	1.0123	1.0267	1.0389
16:3	0.9965	0.9252	0.9875	1.0033	1.0046	1.0189	1.0310
16:4	0.9989	0.9181	0.9799	0.9957	0.9970	1.0111	1.0232
17:0	1.0091	0.9369	**1.0000**	1.0160	1.0173	1.0318	1.0440
17:1	1.0019	0.9302	0.9929	1.0088	1.0101	1.0244	1.0366
18:0	**1.0000**	0.9284	0.9910	1.0068	1.0082	1.0225	1.0347
18:1	0.9932	0.9221	0.9842	**1.0000**	1.0013	1.0155	1.0276
18:2	0.9865	0.9195	0.9776	0.9933	0.9946	1.0087	1.0207
18:3	0.9797	0.9096	0.9709	0.9864	0.9877	1.0017	1.0137
18:4	0.9730	0.9034	0.9642	0.9797	0.9809	0.9949	1.0067
19:0	0.9919	0.9209	0.9830	0.9987	**1.0000**	1.0142	1.0263
20:0	0.9846	0.9141	0.9757	0.9913	0.9926	1.0067	1.0187
20:1	0.9785	0.9085	0.9697	0.9852	0.9865	1.0005	1.0124
20:2	0.9724	0.9028	0.9636	0.9791	0.9803	0.9943	1.0061
20:3	0.9663	0.8971	0.9576	0.9729	0.9742	0.9880	0.9998
20:4	0.9603	0.8916	0.9516	0.9669	0.9681	0.9819	0.9936
20:5	0.9542	0.8859	0.9456	0.9607	0.9620	0.9757	0.9878
21:0	0.9780	0.9080	0.9692	0.9847	0.9860	**1.0000**	1.0119
22:0	0.9720	0.9024	0.9632	0.9787	0.9799	0.9939	1.0057
22:1	0.9664	0.8972	0.9577	0.9730	0.9743	0.9881	0.9999
22:2	0.9609	0.8921	0.9522	0.9675	0.9687	0.9825	0.9942
22:3	0.9554	0.8870	0.9468	0.9619	0.9632	0.9769	0.9885
22:4	0.9499	0.8819	0.9413	0.9564	0.9577	0.9713	0.9828
22:5	0.9443	0.8767	0.9358	0.9508	0.9520	0.9655	0.9770
22:6	0.9388	0.8716	0.9303	0.9452	0.9465	0.9599	0.9713
23:0	0.9665	0.8973	0.9578	0.9731	0.9744	0.9882	**1.0000**
24:0	0.9615	0.8927	0.9528	0.9681	0.9694	0.9831	0.9948
24:1	0.9564	0.8879	0.9478	0.9629	0.9642	0.9779	0.9896

The symbology represents the principal chain of the FAME in question. The following atomic masses were used: C = 12.0110; H = 1.0079; O = 15.9994. FAMEs positional and geometrical isomers, as well as the ramified methyl esters, possess the same F_{CT} since they show the same C number.

Table 2. Theoretical correction factor (F_{CT}) for FAMEs

COO^- group (molar mass=44.0098g) from FAMEs shows negligible response in FID and, only considering the active carbons (C*) in FAMEs molecules. The following calculations may be done:

Methyl stearate = 18 C^* × 12.0110 = 216.1980;

Mehtyl decanoate = 10 C* × 12.0110 = 120.1100.

Thus, the relative percentage for every FAME will be:

Methyl stearate 298.5080 (100%) and 216.1980 (% of C*) = 72.4262% of C*;

Mehtyl decanoate 186.2936 (100%) and 120.1100 (% of C*) = 64.4735% of C*.

Dividing methyl stearate`s C* percetage by methyl decaniate`s C* percetage:

F_{CT} of methyl decanoate = 72.4262% / 64.4735% = 1.1233.

Thus, the value of 1.1233 must be used as FCT in the quantitative determination of methyl decanoate (10:0Me) to correct FID`s differential response with methyl stearate as IS. Values for other FAMEs are expressed in Table 2.

It is possible to obtain, in a simplified manner, new values of FAEs F_{CT} through the use of another internal standard. Just use the obtained values for methyl stearate as IS (first column-Table 2) and values of the new IS in question. For example, to determine the new F_{CT} value of alpha-linolenic acid methyl ester (18:3Me) using methyl tricosanoate (23:0Me) as IS, just divide 18:3Me F_{CT} by 23:0Me F_{CT}, according to F_{CT} values from Table 2, the calculation is shown in equation 1.

$$F_{CT} \text{ of } 18{:}3 = F_{CT} 18{:}3 \,/\, F_{CT} 23{:}0 = 0.9797 \,/\, 0.9665 = 1.0137 \quad (1)$$

Table 2 shows F_{CT} for several FAMEs with different internal standards. Every entry from Table 2 is in accordance with the values published in the literature.

4.3. Experimental correction factor (F_{CE}) determination

Visentainer (2012) determined FID's experimental correction factors (F_{CE}) using a mixture of several FAME's with known amounts of methyl tricosanoate as IS, and from masses and percentual areas of FAME's and IS, the F_{CE} for every FAME was determined. Table 3 shows the average values with the respective standard deviations.The following equation is used for F_{CE} determination:

$$F_{CE} = A_P \times M_x / M_P \times A_x \quad (2)$$

where: A_P = standard area; M_x = FAME mass; M_P = standard mass; A_x= FAME mass.

http://dx.doi.org/10.5772/57333

	Response factor		Coefficient ©
FAMEs	**Experimental (F_{CE})[a]**	**Theoretical (F_{CT})[b]**	**$C = F_{CE} / F_{CT}$**
12:0	1.0535 ± 0.0095	1.1144	0.9454
14:0	1.0653 ± 0.0093	1.0803	0.9861
16:0	1.0491 ± 0.0092	1.0546	0.9948
18:0	1.0282 ± 0.0094	1.0347	0.9937
18:1n-9	1.0329 ± 0.0098	1.0276	1.0052
18:2n-6	1.0524 ± 0.0189	1.0207	1.0311
18:3n-3	1.0505 ± 0.0168	1.0137	1.0363
20:0	1.0274 ± 0.0083	1.0187	1.0085
20:4n-6	1.0484 ± 0.0198	0.9936	1.0552
20:5n-3	1.0443 ± 0.0239	0.9878	1.0572
22:0	0.9905 ± 0.0092	1.0057	0.9849
22:6n-3	1.0442 ± 0.0278	0.9713	1.0751
23:0	1.0000 ± 0.0000	1.0000	1.0000
24:0Me	0.9874 ± 0.0104	0.9948	0.9926

[a]Average values ± estimation of standard deviation (n = 6), [b]Visentainer *et al.*, 2007. FAMEs = fatty acid methyl esters.

Table 3. FCT and FCE in relation to the internal standard methyl tricosanoate (23:0Me)

In a general manner, the coefficients © between F_{CE} and F_{CT} of saturated FAMEs (Table 3) are closer to 1.000 than that from polyunsaturated FAMEs. Thus, it is recommended the use of a saturated FAME as IS. Coefficients with values equal to 1.000 are preferred, because they indicate that the equipment is working with optimized conditions. A very high coefficient for a saturated fatty acid means that the GCph is not optimized, or there are some inconsistencies with the employed method. It is also important to remind that, every time as possible, fresh standards with proper preparation and storage procedures should be used.

4.4. Determination of fatty acid concentration

For determination of FAME mass in milligrams, equation 2 should be used. However, in order to increase the degree of accuracy in fatty acid/FAME quantification, it is recommended to use F_{CT}. The instrumental and chemical parameters must be optimized to minimize possible errors from these sources. It has been reported that, due to oxidative instability of polyunsaturated fatty acids (PUFAs), it is virtually impossible to obtain and keep standards of this kind with high degrees of purity, and they recommend the use of F_{CT} as a good approach for PUFAs analysis. Transforming equation 2 in function of a mass from a fatty acid X, equation 3 is obtained:

$$M_x = M_P \times A_x \times F_{CT} / A_P \quad (3)$$

where: M_x = FAME mass; M_P = standard mass; A_x= FAME area; F_{CT} = theoretical correction factor; A_P = standard area.

4.5. Final equation and FAME conversion factor for foods

In food analysis, results must be expressed in mass of fatty acids, unlike biodiesel, in which results must be expressed in FAME. Thus, sample mass (M_A) and the conversion factor of methyl ester to fatty acid (F_{CEA}) must be added to equation 3, and this new equation (shown below) is called "final equation for determination of fatty acids in mg/g of oil or fat":

$$M_x = M_P \times A_x \times F_{CT} / A_P \times M_A \times F_{CEA} \quad (4)$$

where: M_X = mass of fatty acid in mg/g of oil or fat; M_P = internal standard mass in mg; A_X = FAME area; F_{CT} = theoretical correction factor; A_P = internal standard area; M_A = sample mass (oil or fat) in g; F_{CEA} = conversion factor of methyl ester to fatty acid.

The F_{CEA} is determined through division of a FAME molecular mass (MM) by the MM of its corresponding fatty acid, according to equation 5:

$$F_{CEA} = \text{MM of FAME / MM of the corresponding fatty acid} \quad (5)$$

Fatty acids	MM of FA	MM of FAME	F_{CEA}
Tetradecanoic- 14:0 ($C_{14}H_{28}O_2$)	228	242	1.061
Hexadecanoic- 16:0 ($C_{16}H_{32}O_2$)	256	270	1.055
Octadecanoic- 18:0 ($C_{18}H_{36}O_2$)	284	298	1.049
9-Octadecenoic- 18:1n-9 ($C_{18}H_{34}O_2$)	282	296	1.050
9,12-Octadienoic- 18:2n-6 ($C_{18}H_{32}O_2$)	280	294	1.057
9,12,15-Octadecatrienoic- 18:3n-3 ($C_{18}H_{30}O_2$)	278	292	1.050
Eicosanoic- 20:0 ($C_{20}H_{40}O_2$)	312	326	1.045
5,8,11,14-Eicosatetraenoic- 20:4n-6 ($C_{20}H_{32}O_2$)	304	318	1.046
5,8,11,14,17-Eicosapentaenoic- 20:5n-3 ($C_{20}H_{30}O_2$)	302	216	1.044
Docosanoic- 22:0 ($C_{22}H_{44}O_2$)	340	354	1.041
4,7,10,13,16,19-Docosahexaenoic- 22:6n-3 ($C_{22}H_{32}O_2$)	328	342	1.042

The following atomic masses were taken in account: H = 1; C = 12 e O = 16. MM = molecular mass, FA = fatty acid, FAME = fatty acid methyl ester.

Table 4. Conversion factor of methyl ester to fatty acid F_{CEA}

The FAME of a determined fatty acid possesses methyl group (CH_3) in substitution of an atom of hydrogen. Thus, the methyl ester will show greater response in FID in relation to its corresponding FA, because the CH_3 contributes to increase C* number. Table 4 shows MM and FCEA values for some methyl esters and their respective FAs.

In case of biodiesel determination (FAMEs), F_{CEA} must be equal to 1, because in biodiesel case results are expressed in FAME and not in fatty acids.

5. Exemplifying the fatty acid quantification using gas chromatography

Experimental design, sample preparation, interpretation and processing of generated data are the intrinsic components of a successful GC analysis. In order to assist especially beginner scientists to get started smoothly, the fatty acid quantification of a sardine sample will be demonstrated in a practical manner, using all of the previously described methodologies: normalization, internal standard calibration, use of F_{CT}, F_{CE} and F_{CEA} [3].

5.1. Lipidic extraction (total lipids)

There are several methods of extraction and determination of total lipids in oils and fats from foods. It must be considered that some fatty acids derived from them are unstable, especially the unsaturated ones. The steps that come before FAs analysis must be attended with care, in order to protect the lipidic constituents. Such measures include: (1) the avoidance of metal utensils during sample preparation, (2) packing under vacuum and in a dark place to reduce interactions of light and oxygen with analytes, (3) storing samples in frozen state and (4) use of lipid extraction methods at room temperature.

Among methods that may be used for extracting lipids without system heating, the Bleigh and Dyer method (1959) [10] is one of the most recommended. In this method, three solvents are employed: chloroform ($CHCl_3$), methanol (H_3COH) and water (H_2O). The obtained extract is more reliable for evaluating FAs compositions.

5.2. Bligh and Dyer (1959) method – Simplified

Transfer 100g of homogenized sardine meat to a becker with capacity of 500mL. Add a precise amount of 100mL of chloroform and 200mL of methanol. The contribution of food's water content must be 80%, in order to keep the proportions of chloroform:methanol:water at (1:2:0.8). If it is needed, correct the water content. Stir vigorously during 5 minutes. Add 100mL of chloroform and 100mL of distilled water. Return to vigorous stirring for 5 minutes. The final proportion of solvents, as mentioned above, must be of 1chloroform:2methanol:0.8water. Filter in a Büchner funnel, transfer the filtrate to a separatory funnel and let the aqueous and organic phases to separate. Collect the lower layer (organic phase-chloroform), transfer it to a previously weighed flat-bottomed flask, remove the chloroform in excess through rotary evaporation at 33-35^{0}C and determine the remaining total lipids through gravimetry. After such determination, transfer the total lipids from the flask to a amber-type recipient, and cover the new recipient into aluminium foil. Samples must be stored at -18 °C.

5.3. Preparation of internal standard solution

The first step of analysis is to prepare a solution of the ester that will be used as internal standard. Regardless of FA type, the steps described below always will be the same.

In this moment the analyst must know, through bibliographic research, which FAs will be present in this sample, because it must not contain the ester that will be used as IS. For example, animal fat possesses 17:0Me and 19:0Me FAs, thus these two compounds cannot be used for IS calibration. In this case, 23:0Me is used instead.

The internal standard must be of analytical grade (high purity degree) in order to avoid presence of impurities that might compromise future results.

The IS mass must be measured in an analytical balance and transferred to a volumetric flask (for instance, 50 mL for 50 mg IS). A bit of solvent should be added to the flask. Alkanes such as n-hexane, n-heptano and iso-octane are recommended to be used as solvents, although other alkanes may also be employed. In this case, iso-octane is chosen for use. The final solution should be stored in amber flask and under refrigeration. Other volumes might be prepared, according to the number of samples to be analyzed.

5.4. Transesterification of total lipids

After extraction of total lipids from the sample, they must be derivatized for further FAs analysis through GC. There are several methods for this purpose. For the example of this chapter, the most recommended methodology for fish samples is the one reported by Joseph and Ackman (1992) [4]. This method originated from a collaborative study which involved 21 international laboratories.

In a screw-capped glass tube, 1 mL of the previously prepared IS solution is added. Soon after, solvent is removed by a gentle flow of gaseous N_2. Then, approximately 25 mg of total lipids are weighed in the same tube, and to it 1.5 mL of a methanolic solution of NaOH 0.5 mol/L is added. The entire system is double-boiled to 100 °C for five minutes, and cooled to room temperature.

After cooling, 2.0 mL of solution which corresponds to 12% of BF_3 in methanol is added, and the system is once again double-boiled to 100 °C for thirty minutes and rapidly cooled with running water to room temperature. Immediately after this step, 1 mL of iso-octane is added to the tube, followed by vigorously shaking for 30 seconds and, finally, 5 mL of a saturated NaCl solution is added. The esterified sample is left to rest in a fridge, to allow a better separation of phases. The supernatant is removed and transferred to a 5 mL amber flask. An additional iso-octane amount is further added to the tube. After shaking, the new supernatant is removed and united with the previous fraction in the amber flask, and this final mixture is concentrated to a final volume of approximadetely 1 mL, with the help of gaseous N_2.

5.5. Chromatographic analysis

In the sequence, a GCph is needed for a proper chromatographic run. For the example of this book, 2μL of sample were injected and separated in a GCph from Thermo brand, model 3300, equipped with a flame ionization detector, automatic injector and a capillary column of fused silica CP-7420 SELECT-FAME (100 % bonded cyanopropyl, 100m length, 0.25mm internal diameter and 0.39 μm of stationary phase). Injector temperature was 230°C. Initially, the column temperature was maintained at 165°C for 18 minutes. Then, it was raised to 235°C, at a rate of 4°C/min. The flow rates for the carrier (H_2), auxiliary (N_2) and detector flame (H_2 and synthetic air) gases were 1.2 mL/min, 30 mL/min, 30 mL/min and 300 mL/min, respectively. Sample split ratio was 1/80. Figure 1 is an illustrative chromatogram for the sardine oil sample that was esterified. It can be observed in the chromatogram a peak regarding solvent, which must be removed. After this removal, only the peak areas from esters will compose the chromatogram.

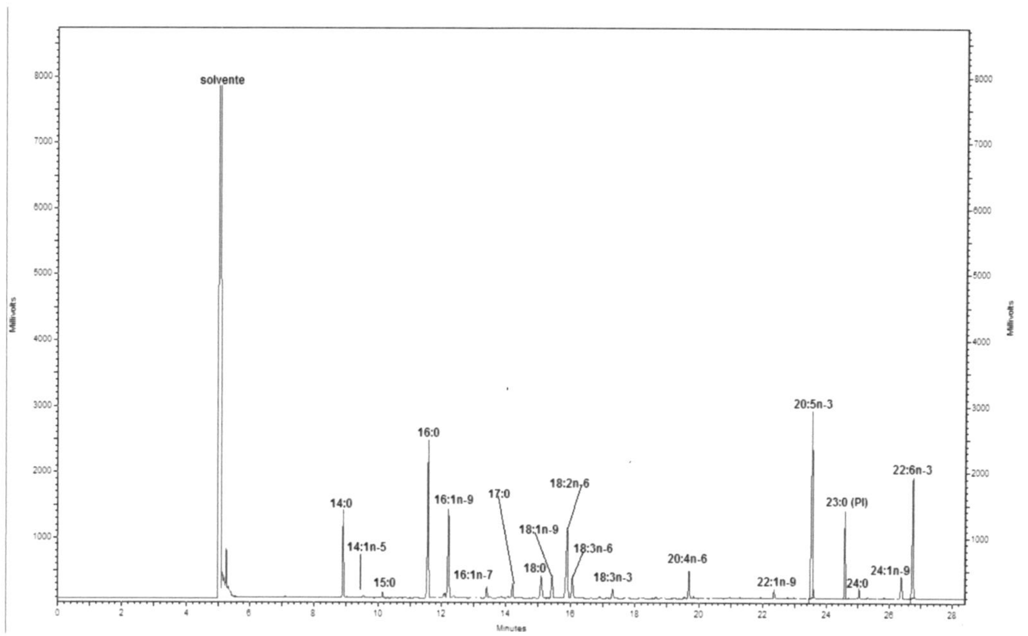

Figure 1. Ilustrative chromatogram of the obtained esters from sardine oil sample with a known amount of internal standard (23:0Me).

Table 5 shows the obtained areas for the different esters (these areas are provided by the software which is used for peak integration).

The values in percentage area (%) were calculated through the normalization method, which will be detailed below:

Assuming to calculate the percentage of 22:6n-3 (DHA) in sample, the following formula can be used:

Methyl esters (ME)	Area	Area in percentage (%)
14:0	21169396	7.32
14:1n-5	1061405	0.37
15:0	3288731	1.14
16:0	80167690	27.71
16:1n-9	3076733	1.06
16:1n-7	13561700	4.69
17:0	3607101	1.25
18:0	16666471	5.76
18:1n-9	23694148	8.19
18:2n-6	6308497	2.18
18:3n-6	4883567	1.69
18:3n-3	1984215	0.69
20:4n-6	4888882	1.69
22:1n-9	1774930	0.61
20:5n-3	28582886	9.88
23:0 (IS)	10889779	-
24:0	2684844	0.93
24:1n-9	3789879	1.31
22:6n-3	68164925	23.56
Total	289356000	100.00

Table 5. Methyl esters which were identified in the chromatogram as well as their respective areas and relative area percentages.

X [%] = Area of 22:6n-3 (DHA) × 100 / Total area (with exception of IS

= 23.56 [%] of DHA.

The procedure is repeated for every peak, and the final sum of percentages is equal to 100%.

The problem of this method is that it considers that every compound present in sample was eluted through the column and detected. This is not always true because some compounds might stay retained on the column. Besides, as mentioned before, the detector must have the same response for every compound. However, this does not happen with FID, because it shows differential responses between the different carbonic chains of methyl esters. Thus, it is not correct to affirm that, in this sardine oil sample, the DHA concentration corresponds to 23.56%.

In order to correct this problem, it is recommended to use conversion factors, as mentioned before. Thus, now it will be exemplified for the same fatty acid (DHA), which would be its percentage after using the correction factor for fish.

For a fish meat sample with 3% of total lipids, the conversion factor is 0.89 (Table 1). Thus, the % of DHA in sample is obtained:

% DHA = (% area of DHA) × (conversion factor).

= (23.56) x (0.89) = 20.97%.

This calculation may be repeated for the remaining FAs, through use of the equation above, but with their respective values of area percentage and correction factors.

This method is very practical and easily done to express results in amount of FAs *per* amount of food. However, it has some critics:

1. In the application, FID's differential response for the different FAs are not taken tinto account;
2. The different total lipid fractions are dependent of many variables, and the tabulated values might not correspond to the real values. A way of minimizing this problem is the empirical determination of the desired fractions;
3. In many foods, total lipids were not fractionated. In this case, the percentage of each fraction must be determined, as mentioned in item II. However, a proper method for fractionating should be employed;
4. Calculations involve mass percentage, through the use of normalization method. Thus, all the sample constituents must be present.

Although the FA quantification with use of internal standard calibration is labor-intensive, especially in relation to its implementation in a laboratory, the obtained results compensate for the spent time, because they are reliable and of easy interpretation, allowing comparison of composition with works which use modern quantification methods.

For this same sardine oil sample, the percentage of DHA will be calculated with this method:

Knowing that DHA area from the chromatogram (Figure 1) is 68164925, in this analysis a 23:0Me 1 mg/mL solution was used as PI, with application of 1 mL (mass 1mg and area of 10889779) from this, the 0.024g of total lipids were used for esterification, the F_{CT} for DHA (Table 3) is 0.97 and the F_{CE} for DHA is 1.04. Thus, through the use of the equation cited below:

$$M_{DHA} = A_x \times M_p \times F_{CT} / A_p \times F_{CE} \times M_A \tag{6}$$

Where:

M_x = Mass of DHA in mg/g of total lipids; M_P = Mass of internal standard in milligrams; M_A = Mass of total lipids in grams; A_X = Area of DHA; A_P = Area of internal standard; F_{CT} = theoretical correction factor; F_{CEA} = conversion factor of methyl ester to fatty acid.

Thus :

M_{DHA} = (68164925).(1).(0.97)/(10889779).(1.04).(0.027)

Finally:

M_{DHA} = 215.61 mg/g of total lipids.

Performing the same calculation for the remaining FAs, it can be obtained the composition of the esterified sardine oil.

Fatty acids	GC-FID			GC-MS
	N (%)	FC (%)	PI	PI
14:0	7.32	6.51	73.29 ± 0.034	73.33 ± 0.031
14:1n-5	0.37	0.33	3.64 ± 0.014	3.62 ± 0.013
15:0	1.14	1.01	11.31 ± 0.017	11.02 ± 0.016
16:0	27.71	24.66	273.95 ± 0.114	271.14 ± 0.103
16:1n-9	1.06	0.95	10.41 ± 0.010	10.78 ± 0.009
16:1n-7	4.69	4.17	45.91 ± 0.016	45.20 ± 0.014
17:0	1.25	1.11	12.13 ± 0.015	9.11 ± 0.014
18:0	5.76	5.13	55.66 ± 0.018	54.91 ± 0.016
18:1n-9	8.19	7.29	79.05 ± 0.032	82.18 ± 0.029
18:2n-6	2.18	1.94	20.84 ± 0.029	17.26 ± 0.026
18:3n-6	1.69	1.50	15.98 ± 0.256	15.28 ± 0.230
18:3n-3	0.69	0.61	6.49 ± 1.287	6.21 ± 1.158
20:4n-6	1.69	1.50	15.81 ± 1.992	14.83 ± 1.792
22:1n-9	0.61	0.55	5.71 ± 0.422	5.91 ± 0.380
20:5n-3	9.88	8.79	91.43 ± 0.288	87.50 ± 0.259
23:0 (PI)	-	-	-	-
24:0	0.93	0.83	8.71±0.139	8.08 ± 0.126
24:1n-9	1.31	1.17	12.17 ± 0.144	11.11 ± 0.130
22:6n-3	23.56	20.97	215.61 ± 0.302	214.57 ± 0.272
Total	100.00	89.02	958.10	942.04

N = values determined through the normalization method; FC = values determined through the correction factor method ; PI = values determined through internal standard calibration method with FID correction factors.

Table 6. Obtained values for sardine oil quantification in GC-FID and GC-MS.

For comparison effects, 1μL of the same esterified sardine oil sample was injected in GC-MS. In the example of this book, samples were separated in a GC-MS from Thermo brand, model Focus DSQ II, equipped with automatic injector, and a capillary column DB-5 (5% phenyl and 95% methylpolisiloxane, 30m length, 0.25mm internal diameter and 0.25 μm of stationary phase). Initially, the column temperature was maintained at 165°C for 4 minutes. It was then raised to 200°C, at a rate of 6°C/min, and kept at this temperature for 5 minutes. After this period, it was once again raised to 235°C at a rate of 7°C/min and maintained for 5 minutes. Finally, it was raised for the last time to 290°C at a rate of 20°C/min and kept at this temperature for 8 minutes. The flow rate for carrier gas (He) was 1.0 mL/min. Injector temperature was 240°C. Sample split ratio was 1/20. Data from MS detection was obtained in full scan mode, with range of masses 50-650 m/z and 0.8170 scans per second. Temperature of ionization source was 250°C. Since the MS detector does not show differential response between the different carbonic chain of FAMEs, a very reliable quantification is obtained, and the results of this analysis can be compared with the ones from GC-FID as given in Table 6.

6. Conclusion

The illustrations and results shown here indicate that it is possible to increase accuracy in expression of fatty acids (in foods) or FAEs (in biodiesel), especially when compared with the area normalization method, which is limited and frequently used in an erroneous manner. Besides, results of analyses using internal standard calibration and correction factors, in a sardine oil sample, showed that the expression of concentration results, in mass of fatty acid *per* oil mass, must be used whenever possible, because in this quantification only relevant peaks from a chromatogram are taken into consideration. Thus, the obtained results are reliable and easily interpreted, allowing quantitative comparisons of fatty acids/FAEs in foods/biodiesel. The support of such results were obtained upon comparison of the same sample using GC-FID and GC-MS, and the results were indistinguishable, demonstrating that FID may be used for fatty acid analysis, as long as the proper correction factors are applied. Thus, using FID for such quantifications is highly recommended and, comparing to MS, it has additional advantages because it is cheaper and easier to operate, while keeping good sensitivities.

Author details

Jesuí Vergilio Visentainer*, Thiago Claus, Oscar Oliveira Santos Jr, Lucas Ulisses Rovigatti Chiavelli and Swami Arêa Maruyama

*Address all correspondence to: swamimaruyama@yahoo.com.br

Maringá State University, Department of Chemistry, Maringá-PR, Brazil

References

[1] Collins, C. H.; Braga, G. L.; Bonato, P. S.; *Fundamentos de Cromatografia,* Ed. da Unicamp: Campinas, 2006.

[2] Santos, F. J.; Galceran, M. T. Modern developments in gas chromatography-mass spectrometry-based environmental analysis. *Journal Chromatography. A* 2003; 1000, 125-151.

[3] Visentainer, J. V.; Franco, M. R. B.; *Ácidos Graxos em Óleos e Gorduras: Identificação e Quantificação*, 1a ed, Varela: São Paulo, 2006.

[4] Joseph, J. D.; Ackman, R. G.; Capillary column gas chromatography method for analysis of encapsulated fish oil and fish oil ethyl esters: collaborative study. *Journal of AOAC International.* 1992; *75*, 488-506.

[5] Visentainer, J. V. Aspectos analíticos da resposta do detector de ionização em chama para ésteres de ácidos graxos *Quimica. Nova* 2012; 35, 274-279.

[6] Ackman, R. G.; The analysis of fatty acids and related materials by gas-liquid chromatography. *Progress in the Chemistry of Fats and other Lipids*. 1972; *12*, 165-284.

[7] Feng, X.; Liu, X.; Luo, Q.; Liu, B. F. Mass spectrometry in systems biology: an overview. *Mass Spectrometry Reviews* 2008; 27, 635-660.

[8] Holm, T.; Aspects of the mechanism of the flame ionization detector. *Journal of Chromatography A*, 1999; *842*, 221-227.

[9] Ackman, R.; Sipos, J. C.; Flame ionization detector response for the carbonyl carbon atom in the carboxyl group of fatty acids and esters. *Journal of the American Oil Chemists Society* 1964; *16*, 298-305.

[10] Bligh, E. G.; Dyer, W.; A rapid method of total lipid extraction and purification. *Canadian journal of biochemistry and physiology*. 1959, *37*; 911-917.

Chapter 3

Limit of Detection and Limit of Quantification Determination in Gas Chromatography

Ernesto Bernal

Additional information is available at the end of the chapter

http://dx.doi.org/10.5772/57341

1. Introduction

Any year, millions of analyses of any kind are performed around the world, and millions of decisions are made, based on these analyses; have the medicaments, the amount of drug reported in their container?, Can we safely consume this water or these foods?, are these alloys suitable for use in the aircraft construction?, was the driver drunk, when he crashed?, is this sportsman using drugs to enhance his performance?, and if he was punished, there is no doubt, he was using those substances or we are unfair with him; all these questions are answered with the help of a chemical analysis and all have consequences in real life (compensation claims, disease, fines, even prison). Virtually every aspect of society is supported in some way by analytical measurement, consequently, there is a need these analyses would be reliable.

Until the 1970´s, the underlying assumption was that the reports submitted by laboratories accurately described study conduct and precisely reported the study data. Suspicion about this assumption was raised during the review of some studies. Data inconsistencies and evidence of unacceptable laboratory practices came to light [1]. If the result of a test cannot be trusted then it has little value and the test might as well have not been carried out. When a client commissions analytical work from a laboratory, it is assumed that the laboratory has a degree of expertise that the client does not have. The client expects to be able to trust the results reported. Thus, the laboratory and its staff have a clear responsibility to justify the clients trust by providing the right answer to the analytical part of the problem, in other words, results that have demonstrable "fitness for purpose"'[2]. Implicit in this is that the tests carried out are appropriate for the analytical part of the problem that the client wishes solved, and that the final report presents the analytical data in such a way that the client can readily understand it and draw appropriate conclusions. Method validation enables chemists to demonstrate that a method is "fit for purpose" [1]. For an analytical result to be fit for its intended purpose it must be sufficiently reliable that any decision based on it can be taken with confidence. Thus the method performance must be validated and the uncertainty on the result, at a given level of confidence, estimated. Uncertainty should be evaluated and quoted in a way that is widely recognized, internally consistent and easy to interpret. Most of the information required to evaluate uncertainty can be obtained during validation of the method [1].

Since then, several agencies like United States Food and Drug Administration (FDA) [3-5], the International Conference for Harmonization (ICH) [6], the United States Pharmacopeia (USP) [7], the International Standards Organization (ISO/IEC) [8], etc. created working groups to ensure the validity and reliability of the studies. They would eventually publish standards for measurement the performance of laboratories and enforcement policy. Good laboratory practice (GLP) regulations were finally proposed in 1976, being method validation an important part of GLP.

1.1. Method validation

ISO [9] defines validation as the confirmation, via the provision of objective evidence, that the requirements for specifically intended use or application have been met, so method validation is the process of defining an analytical requirement, and confirming that the method under consideration has performance capabilities consistent with what the application requires [2].

Therefore, method validation should be an essential component of the measurements that a laboratory makes to allow it to produce reliable analytical data, in consequence, method validation should be an important part in the practice of all the chemists around the world. Nevertheless, the knowledge of exactly what needs to be done to validate a method seems to be poor amongst analytical chemists. The origin of the problem is the fact that many of the technical terms used in processes for evaluating methods vary in different sectors of analytical measurement, both in terms of their meaning and also the way they are determined [2].

In order to resolve the previous problem, several protocols and guidelines [2, 7, 10-19] on method validation have been prepared. Method validation consists of a series of tests that both proof any assumptions on which the analytical method is based and established and document the performance characteristics of a method, demonstrating whether the method fits for a particular analytical purpose [15]. Typical performance characteristics of analytical methods are applicability, selectivity or specificity, calibration, accuracy and recovery, precision, range, limit of quantification, limit of detection, sensitivity and ruggedness or robustness [2, 7, 10-19], Unfortunately, some of the definitions vary between the different organizations producing confusion among analysts. Some parameters are summarized in Table 1.

It´s not the purpose of this work to define the previous parameters and the procedure to determine them or the strategies to perform a validation, all of them concerned with these issues are invited to consult the following references [2, 7, 10-19].

http://dx.doi.org/10.5772/57341

Parameter	Comment
Specificity	USP, ICH
Selectivity	ISO 17025
Precision	USP, ICH
• Repeatability	ICH, ISO 17025
• Intermediate precision	ICH
• Reproducibility	ICH
Accuracy	USP, ICH, ISO 17025
Linearity	USP, ICH, ISO 17025
Range	USP, ICH
Limit of detection	USP, ICH, ISO 17025
Limit of quantification	USP, ICH, ISO 17025
Robustness	USP; ISO 17025
Ruggedness	USP

Table 1. Parameters for method validation

1.2. Limit of detection

From the previous section, it is clear that despite the efforts to standardize concepts, there is still confusion about some terms in method validation, like selectivity and specificity, ruggedness and reproducibility, accuracy and trueness. Nevertheless, the most troublesome concept of all, in method validation is the limit of detection (LOD). LOD remains an ambiguous quantity on analytic chemistry in general and gas chromatography in particular; LOD's differing by orders of magnitude are frequently found for very similar chemical measurement process (CMP). Such discrepancies raise questions about the validity of the concept of the LOD.

The limit of detection is the smallest amount or concentration of analyte in the test sample that can be reliably distinguished from zero [20]. Despite the simplicity of the concept, the whole subject of LOD is with problems, translating these into the observed discrepancies in the calculation of the LOD. Some of the problems are [15]:

- There are several conceptual approaches to the subject, each providing a somewhat different definition of the limit, and consequently, the methodology used to calculate the LOD derived from these definitions, differ between them.
- LOD is confused with other concepts like sensitivity.
- Estimates of LOD are subject to quite large random variation.
- Statistical determinations of LOD assume normality, which is at least questionable at low concentrations.
- The LOD, which characterizes the whole chemical measurement process (CMP), is mistaken with concepts that characterize only one aspect of the CMP, the detection.

These problems are more prominent in the field of chromatography, where, besides the previous issues, no standard model for the LOD has ever proposed by any recognized organization. Actually, the International Union of Pure and Applied Chemistry (IUPAC) model for LOD determination was chosen for spectrochemical analysis specifically. Thus, chromatographic conditions are usually not taken in consideration to determine the LOD [15, 21].

The main purpose of this paper is to bring some light to these problems. In order to achieve this goal, the different problems behind LOD and limit of quantification (LOQ) are going to be discussed. The different definitions and conceptual approaches to LOD and LOQ, given by different associations [2,7,20, 22], the different models to calculate LOD and LOQ, and the effect of matrix and particularities related to chromatographic techniques on LOD and LOQ calculations are going to be critically reviewed [23-25], aiming at unifying criteria and estimating LOD and LOQ in a more reliable figure of merit in chromatography.

2. Limit of detection and limit of quantification

2.1. Definitions

Since the seminal work of Currie [26], emphasis has been placed on the negative effect it has had, the large number of terms that have been used through the years regarding the detection capabilities of a method (table 2). A wide range of terminologies and multiple mathematical expressions have been used to define the limit of detection concept. These different terms resulted in different ways of calculating the LOD, leading to numerical values that can span over three orders of magnitude, applied to the same measurement process. Some mathematical definitions involved the standard deviation of the blank, some the standard deviation of the net signal, some authors used two sided confidence intervals, while others used one sided intervals, and some authors even used non statistical definitions; what was missing from these authors was a theoretical basis for the concept that led to an operational definition of the term.

To overcome the previous problem, the ISO and the IUPAC developed documents bringing their nomenclature into essential agreement [15, 20, 22]. As the measure of the detection capability of a CMP, the IUPAC recommends the term Minimum Detectable Value (L_D) of the appropriate chemical variable or Detection limit and defines it as the smallest amount or concentration of analyte in the test sample that can be reliably distinguished from zero. And as the measure of the quantification capability of the CMP, the IUPAC recommends the term Minimum quantifiable limit (L_Q) or Quantification limit, which is the concentration or amount below which the analytical method cannot operate with an acceptable precision. ISO definition puts emphasis in the statistics; ISO [2, 22] defines the minimum detectable net concentration as the true net concentration or amount of the analyte in the material to be analyzed, which will lead with probability (1-β) to the conclusion that the concentration of the analyte in the analyzed material is larger that of the blank matrix.

In addition, several organizations have introduced terms with similar meaning to LOD. The US Environmental Protection agency (EPA) uses the term method detection limit (MDL) as the minimum concentration of an analyte that can be identified, measured and reported with 99% confidence that the analyte concentration is greater than zero.

On the other hand, the ICH defines LOD as the lowest amount of analyte in a sample which can be detected but not necessarily quantitated as an exact value and LOQ of an individual analytical procedure as the lowest amount of analyte in a sample which can be quantitatively determined with suitable precision and accuracy [19, 27].

The American Chemical Society (ACS) committee on environmental improvement states LOD as the lowest concentration of an analyte that the analytical process can reliably detect [27].

The conference for drug evaluation and research (CDER) in its document entitled 'Validation of chromatographic method' defines LOD as the lowest concentration of analyte in a sample that can be detected but not necessarily quantitated under the stated experimental conditions [19, 27].

The national association of testing authorities (NATA) defines LOD as the smallest amount or concentration that can be readily distinguished from zero and be positively identified according to predetermined criteria and/or level of confidence, while the lowest concentration of an analyte that can be determined with acceptable precision (repeatability) and accuracy under the stated conditions of the test is the limit of quantification [2, 27].

On the other hand, the AOAC defines limit of detection as the lowest content that can be measured with reasonable statistical certainty and the limit of quantification as the content equal to or greater than the lowest concentration point on the calibration curve [2].

Finally, the USP [7] defines LOD as: the lowest amount of analyte that can be detected but not necessarily quantitated under the stated experimental conditions. Table 2 summarizes these terms, symbols and statistical items reported in the literature [28].

All definitions include terms as reliability, probability, confidence, etc. That implies the use of statistics to calculate them. Some definitions even put inside its text explicitly the degree of reliability and consequently, define it. The others leave the decision to the operator. Some definitions make it clear that they are referring to a CMP and not only to the detection phase of the analysis. Therefore, these terms should not be confused with terms referred only to detection like the Instrumental Detection Limits (IDL) used by EPA. The limit of detection is a parameter set before the measure, and, in other words, is defined a priori; therefore, is not related to the decision whether a measurement detects anything or not. Finally, these terms should not be confused with sensitivity defined as the slope of the calibration curve by IUPAC.

As an example of the wide range of terms used to define detection capabilities of a method used nowadays, it is shown a study performed in 2002 [29], by the American Petroleum Institute (API), to intend to review policies related to analytical detection and quantification limits of ten states of the USA, with particular focus on water quality and wastewater issues in permitting and compliance. Thus, these regulations should follow the EPA recommendations. It was found that every state incorporates detection or quantification terms in its regulations to some extent. Terms referenced are usually defined in the regulations, but not always. The most frequently used terms are detection limit/level, method detection limit (MDL), limit of detection (LOD), and practical quantitation level (PQL). Minimum Level (ML) is the term used by EPA instead of LOQ; it is defined as the concentration at which the entire analytical system must give a recognizable signal and acceptable calibration point. The ML is the concentration in a sample that is equivalent to the concentration of the lowest calibration standard analyzed by a specific analytical procedure, assuming that all of the method-specified sample weights, volumes, and processing steps have been followed. EPA uses other terms like Interim minimum level (IML) which is a term created by the EPA to describe MLs that are based on 3.18 times the MDL, to distinguish them from MLs that have been promulgated. The EPA defines the PQL as: The lowest level that can be reliably achieved within specified limits of precision and accuracy during routine laboratory operating conditions. Another term used by EPA is the alternative minimum level (AML) which can account for interlaboratory variability and sample matrix effects and finally the Interlaboratory Quantification estimate (IQE) a term developed by the American Society for Testing and Materials (ASTM), which is similar to the AML. Table 3 lists detection and quantification terms used by the ten states.

Terms	Signal Domain	Concentration Domain	Reference
Limit of detection	LOD	LOD	IUPAC 1976
		LOD	ACS, 1980
Critical Level	L_C	X_C	Oppenheimer, et al, 1983
	y_C	X_C	Hubaux & Vos, 1970
Critical value	L_C	X_C	Currie, 1995
Method detection limit		MDL	EPA
Limit of guarantee of purity	x_G	c_G	Kaiser, 1966
Limit of identification	x_I	c_I	Long & Winefordner, 1983
Detection level	L_D	X_D	Oppenheimer, et al, 1983
Detection limit	y_D	x_D	Hubaux & Vos, 1970
Minimum detectable value	L_D	X_D	Currie, 1995
Limit of quantification		LOQ	ACS, 1980
	LOQ		Long & Winefordner, 1983
Determination Limit	L_Q	x_Q	Oppenheimer, et al, 1983
Minimum Quantifiable level	L_Q	x_Q	Currie, 1995
Minimum Level		ML	EPA
		PQL, IML	EPA

Table 2. Terms and symbols reported in the literature related to method detection capabilities

http://dx.doi.org/10.5772/57341

Therefore, despite the efforts of various agencies to standardize the terms concerning the detection capability of a measurement process, there are still differences between the various regulations.

2.2. Theory

2.2.1. Hypothesis testing approach

In 1968, Currie published the hypothesis testing approach to detection decisions and limits in chemistry [26]. This approach has gradually been accepted as the detection limit theory.

Currie´s achievement was in recognizing that there were two different questions under consideration when measurements were performed on a specimen under test, and these two questions have different answers. The first question, which was "does the measurement result indicate detection or not?", is answered by performing measurements on the specimen under test then computing an appropriate measure for comparison with a critical decision level. The second question is "what is the lowest analyte content that will reliably indicate detection?" and the answer is defined as the detection limit.

State programs	DL	IDL	IML	LOD	LOQ	MDL	ML	QL	PQL	Other§
				Alabama						
Groundwater soil/ hazardous waste	x			x					x	
Water quality	x		x			x	x			
				California						
Drinking water	x						x			DLR
Groundwater soil/ hazardous waste	x	x		x		x			x	
Health/Safety				x						
Water quality					x	x	x		x	
				Illinois						
Drinking water	x					x			x	MQL, ADL
Groundwater soil/ hazardous waste	x			x		x			x	EDL, EQL, MQL
Laboratories						x				
Pesticides	x					x				MQL
Water quality	x			x	x	x	x	x		

State programs	DL	IDL	IML	LOD	LOQ	MDL	ML	QL	PQL	Other[§]
Lousiana										
Air	x			x	x					
Groundwater soil/ hazardous waste	x			x				x	x	
Water quality	x									LOM, MQL**
New Jersey										
Drinking water						x				
Groundwater soil/ hazardous waste	x	x			x	x		x	x	CRDL
Laboratories	x	x				x				
Water quality	x					x			x	NJQL LOD*
Ohio										
Air					x					
Drinking water	x					x				
Groundwater solid/ hazardous waste	x			x					x	ADL
State programs	DL	IDL	IML	LOD	LOQ	MDL	ML	QL	PQL	Other
Laboratories						X				
Underground storage tanks						X		X		
Water quality						X	X	X	X	ADL
Oklahoma										
Air	X									ADL, MLD, MQL*
Drinking water										MDL*
Groundwater solid/ hazardous waste	X			X					X	
Water quality									X	MQL*
Pennsylvania										
Air										
Drinking water						X			X	

http://dx.doi.org/10.5772/57341

State programs	DL	IDL	IML	LOD	LOQ	MDL	ML	QL	PQL	Other[§]
Groundwater solid/ hazardous waste						X		X	X	EQL
Water quality	X			X		X			X	
				Texas						
Air	X									
Drinking water	X					X			x	
Groundwater solid/ hazardous waste	X			X		X			x	MQL, SQL
Water quality						X				MAL
				Washington						
Air									X	
Drinking water	X									
Groundwater solid/ hazardous waste	X			X	X	X			X	
Water quality	X					X			X	

§ Definitions on section 4

Table 3. Summary of detection and quantification terms used by the states

According to Currie, the IUPAC and ISO [20, 30-31], detection limits (minimum detectable amounts) are based on the theory of hypothesis testing and the probabilities of false positives α, and false negatives β. On the other hand, quantification limits are defined in terms of a specified value for the relative standard deviation. It is important to emphasize that both types of limits are CMP performance characteristics, associated with underlying true values of the quantity of interest; they are not associated with any particular outcome or result. The detection decision, on the other hand, is result-specific; it is made by comparing the experimental result with the critical value which is the minimum significantly estimated value of the quantity of interest. In other words it is used to make a posteriori estimate of the detection capabilities of the measurement process, while the limit of detection is used to make a priori estimate.

In order to explain the detection limit theory, let's assume we have an analytical method with known precision along all its concentration levels and that its results follow a normal distribution, if we test a lot of blanks with the method above, certainly a distribution as in Figure 1 can be obtained.

The blank values will distribute around zero with a standard deviation σ_0. In others words if we measure a blank, we can obtain a result different from zero due to the experimental errors of the measure process. Thus we need to establish a point to differentiate blanks measures from non blank measures. That point is the critical value L_C, that point allows us to determine

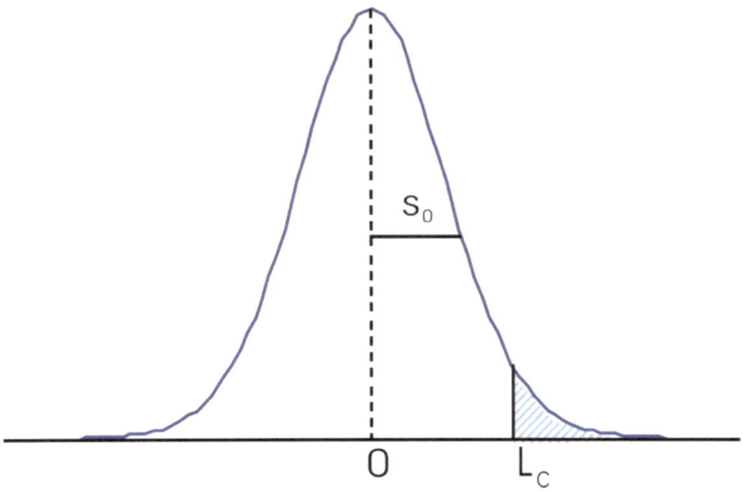

Figure 1. Value distribution around zero and the critical level.

if a signal corresponds to a blank or if the signal is from an analyte, i.e. to make a posterior decision. Nevertheless, in the critical level there is the probability (blue shadow) that a blank can give a signal above the L_C. Therefore we can erroneously conclude that there is analyte, when is not. That probability α, is named type I error or a false positive. The value of α is chosen by the analyst in function of the risk that one wants to take of being wrong. The hypothesis testing theory uses the following definition.

$$\Pr\left(L > L_C | L=0\right) \leq \alpha \tag{1}$$

Where, L is used, as the generic symbol for the quantity of interest. This is replaced by S, when treating net analyte signal and x, when treating analyte concentrations or amount; mathematically, the critical level is given as:

$$L_C = K_\alpha \sigma_0 \tag{2}$$

Where K_α and α, are linked with the one sided tails of the distribution of the blank corresponding to probability levels, 1-α.

Nevertheless, the critical level L_C cannot be used as the limit of detection, because if we measure a series of samples with an amount of analyte equal to the L_C, the results will follow like the blanks a normal distribution around the L_C value (Figure 2). Half of the results would be above the L_C and we conclude that the signal is from an analyte, and half would be below the L_C and consequently, we would think the sample is a blank. Therefore, if we set the L_C as the limit of detection, we would report erroneously half of the results; then, there is the possibility to report

http://dx.doi.org/10.5772/57341

that the analyte is not present in the sample, when it actually is. The probability β, is named type II error or a false negative.

Therefore, if a laboratory cannot accept a 50% of error around the limit of detection, the only alternative to reduce the probability of false negative is to set the limit of detection to a bigger concentration (Figure 3).

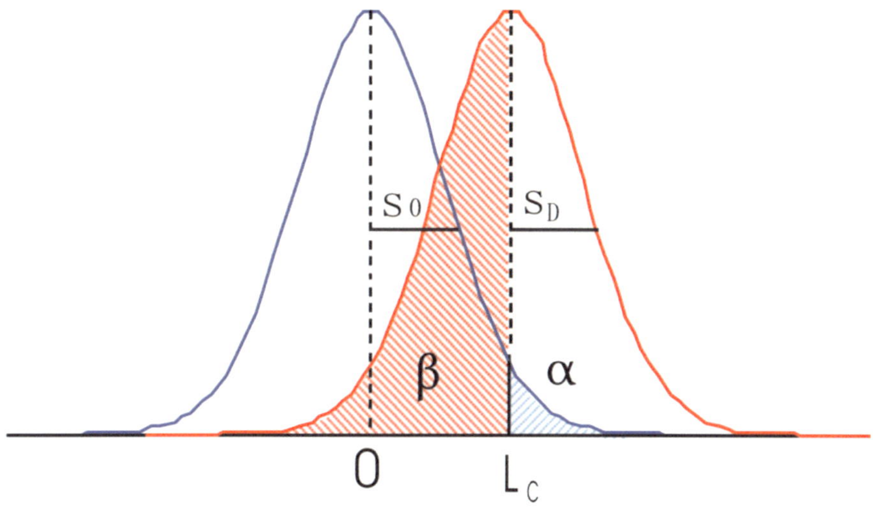

Figure 2. Critical level and limit of detection with probability α. The risk of a false negative is 50%.

Once L_C has been defined, a priori limit of detection L_D, may be established by specifying L_C, the acceptable level β, for the type II error and the standard deviation, σ_D, which characterizes the probability distribution of the signal when its true value is equal to L_D [26]. By the hypothesis testing theory, we obtain the following relation:

$$\Pr\left(L \leq L_C \mid L = L_D\right) = \beta \tag{3}$$

Mathematically, the limit of detection is given as:

$$L_D = K_\alpha \sigma_0 + K_\beta \sigma_D \tag{4}$$

If equation 2 is substituted in equation 4, we obtain:

$$L_D = L_C + K_\beta \sigma_D \tag{5}$$

Where K_β and β, are linked with the one sided tails of the distribution of the limit of detection corresponding to probability levels, 1-β. Finally, the defining relation for the limit of quantification (L_Q) is:

$$L_Q = K_Q \sigma_Q \quad (6)$$

Where K_Q=1/RSD_Q, and σ_Q equals the standard deviation of L when L= L_Q.

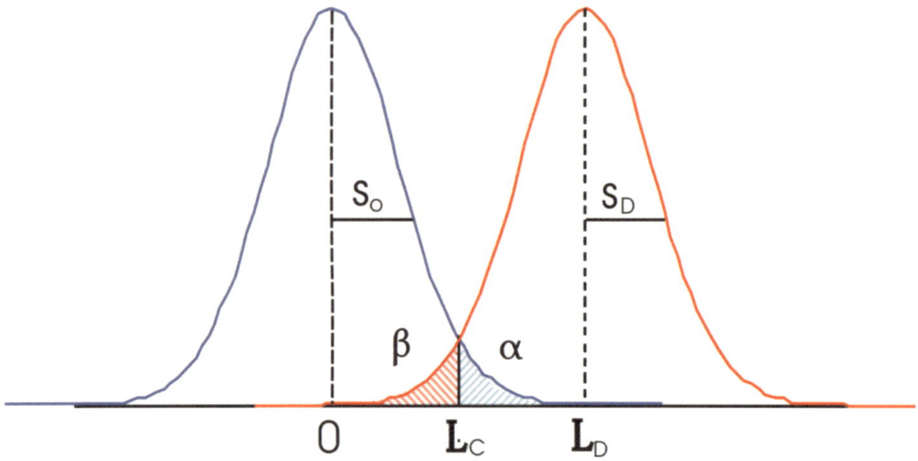

Figure 3. Critical level and limit of detection with α=β.

Summarizing, the levels L_C, L_D and L_Q, are determined by the error structure of the measurement process, the risks α and β and the maximum acceptable relative standard deviation for quantitative analysis. L_C is used to test an experimental result, whereas L_D and L_Q refer to the capabilities of the measurement process itself. The relations among the three levels and their significance in analysis are shown in Figure 4.

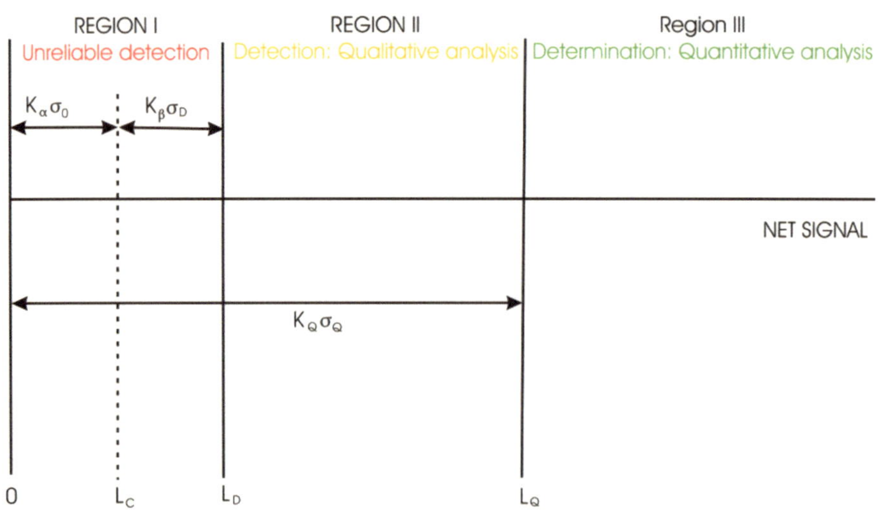

Figure 4. The three principal analytical regions [26].

α, β and K_Q can be chosen by the analyst according to the detection and quantification needs; of particular interest is when α=β, and σ=constant, in that circumstances $K_\alpha = K_\beta = K$, and $\sigma_{D=}\sigma_0$

$$L_C = K\sigma_0 \quad (7)$$

$$L_D = L_C + K\sigma_D = K(\sigma_0 + \sigma_D) \quad (8)$$

And because σ=constant, $\sigma_0 = \sigma_D = \sigma$, relation 7, can be written as:

$$L_D = 2K\sigma = 2L_C \quad (9)$$

In this particular case the limit of detection is twice the critical level. The values of α and β recommended by IUPAC and ISO regulations are 0.05 each, which gives a value of K= 1.645, and since σ=constant, it can choose either the value of σ_0 or σ_D, because L_D is not known, in equation 9 the value of σ_0 is used. Under these circumstances (normal distribution, constant standard deviation and parameters α=β=0.05) the following expressions are obtained [20, 26, 30-31].

$$L_C = K\sigma_0 = 1.645 \quad \sigma_0 \quad (10)$$

If σ_0 is estimated by s_0, based on ν degrees of freedom, K can be replaced by Student's t, resulting in equation:

$$L_C = t_{1-\alpha,\nu} s_0 \quad (11)$$

For α=0.05 and four degrees of freedom, equation 12 is obtained.

$$L_C = 2.132 \quad \sigma_0 \quad (12)$$

Following the procedure developed in the previous paragraphs, equivalent equations are obtained for the limit of detection.

$$L_D = L_C + K\sigma_D = 3.29 \quad \sigma_0 \quad (13)$$

If L_C employs an estimate s_0, based on ν degrees of freedom, then K can be replaced by $\delta_{\alpha,\beta,\nu}$, the non centrality parameter of the non central t distribution. If α=0.05 and four degrees of freedom are used, equation 14 is obtained.

$$L_D = \delta_{\alpha,\beta,\nu}\sigma_0 = 2t_{1-\alpha,\nu}\sigma_0 = 4.26 \quad \sigma_0 \quad (14)$$

The ability to quantify is expressed in terms of the signal or analyte that will produce estimates having a specified relative standard deviation (RSD), commonly 10%. That is

$$L_Q = K_Q \sigma_Q = 10\sigma_0 \tag{15}$$

Where, L_Q is the limit of quantification, σ_Q, the standard deviation at the limit of quantification and K_Q is the multiplier whose reciprocal equals the selected RSD; the IUPAC default value is 10. It's possible to transform all those expressions from the signal domain to concentration domain, and vice versa through the slope of the calibration curve.

The above relations represent the simplest possible case, based on restrictive assumptions. Actually, some of them are questionable as the assumption of normality in the blank measures and that σ is constant along the region of the critical level and limit of detection. They must not be taken as the defining relations for detection and quantification capabilities; being the defining relations, equation one, three and six for the critical level, the limit of detection and the limit of quantification respectively. Finally, for chemical measurement at least, the Fundamental contributing factor to the detection and quantification performance characteristics is the variability of the blank.

Currie's hypothesis testing schema, in spite of being theoretically solid, is very broad in scope, being independent of noise, the methodology to perform the measurements, conditions, etc. In fact his schema does not even have a connection to the calibration curve methodology or even the substance that would be measured.

A particular problem for the calculation of the limit of detection within the field of gas chromatography is the calculation of the deviation of the blank. It has been suggested to use the measurement of 20 blanks and calculate its standard deviation. Other authors suggest measuring the noise at the baseline of one chromatogram in a region near the analyte peak [21]. Nevertheless, questions arise about the sets of the integration parameters, the region of the baseline which should be used to calculate the blank variability and the presence of interfering substances. This makes the determination of s_0 subjective and highly variable and has a major drawback in using the IUPAC definition in dynamic systems such as chromatography [23]. In order to introduce the calibration curve in the limit of calculation, other approaches have been developed to calculate the detection capabilities of a CMP.

2.2.2. Hubaux- Vos approach

In 1970, Hubaux and Vos suggested how Currie's schema could be implemented with calibration curve methodology based on CMP, with homoscedastic (σ=contant), Gaussian noise, ordinary least square processing of the calibration curve data and ordinary least square prediction intervals [32]. Since then, Hubaux and Vos' treatment has generally been assumed to be fundamentally correct.

Hubaux and Vos made a series of assumptions to develop their approach, It was assumed that the standards of the calibration curve are independent, that the deviation is constant through

http://dx.doi.org/10.5772/57341

the calibration curve, the contents of the standards are accurately known and above all, the signals of all the points in the calibration curve have a Gaussian distribution (Figure 5).

Then Hubaux and Vos drew two confident limits on either sides of the regression line with a priori level of confidence (Figure 5), $1-\alpha-\beta$. The regression line and its two confident limits can be used to predict with a $1-\alpha-\beta$ probability, likely values for signals. The confidence band can be used in reverse, for a measured signal y of a sample of unknown content, it is possible to predict the range of its content (x_{max}-x_{min}) (Figure 5).

To our subject, a signal equal to y_C is of interest (Figure 5), where the lower limit of content is zero. Signals equal or lower than y_C have a probability bigger than α due to a blank, and hence cannot be distinguished from a blank signal. y_C is the lowest measurable signal and therefore corresponds to the critical level L_C of Currie. More exactly y_C is an estimate of L_C. Because this limit concerns signals, it is used to posteriori decisions.

If we trace a line from y_C to the lower confident limit and then to the x axis (Figure 5), the value x_D can be obtained, which is the lowest content it can be distinguished from zero. This value is inherent to the CMP and can be used as a priori limit, thus is equivalent to the limit of detection L_D of Currie. It is important to clarify that the regression line and its confident limits are estimates of the real values. Consequently, the values of y_C and x_D are estimates too [32].

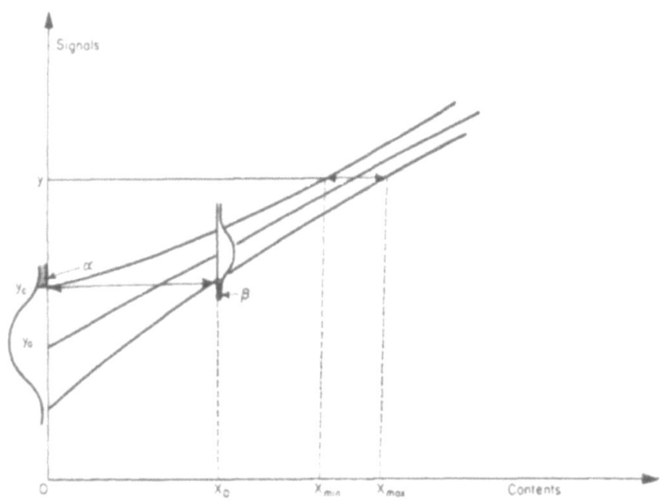

Figure 5. The linear calibration line, with its upper and lower confidence limits. y_c is the decision limit and x_D the detection limit [33].

One serious problem with the Hubaux-Vos approach is the non-constant widths of the prediction interval which contradicts the assumption of homoscedasticity; another problem is, because y_C and x_D are estimates, this method requires the generation of multiple calibration curves to calculate the mean of y_C and x_D.

2.2.3. Propagation of errors approach

In the hypothesis testing approach, the value of the limit of detection L_D depends only on the variability of the blanks σ_0. The propagation of errors approach considers the standard deviation of the concentration s_X. This value is calculated by including the standard deviations of the blank s_0, slope s_m, and intercept s_i, in the equation [25]. The contribution of the variability of slope, blank and intercept to the variability of x is expressed by the formula:

$$s = \frac{\left[s_o^2 + s_i^2 + \left(\frac{1-y}{m} \right)^2 s_m^2 \right]^{1/2}}{m} \tag{16}$$

The standard deviation of the concentration is equal to s_D, and because it was assumed that, the standard deviation is constant along the region of interest s_0=s_D=s, it can substitute s_0 in any of Currie relations. If we substitute equation 16 in equation 13, and we assume the blank's signal is set to zero the following relation is obtained.

$$L_D = \frac{k \left\{ s_0^2 + s_i^2 + \left[\left(\frac{i}{m} \right)^2 s_m^2 \right] \right\}^{1/2}}{m} \tag{17}$$

Where K is a constant related to the degree of error, the analysts assume. The mathematical expressions for s_0, s_i and s_m, can be found in [25] and publications specialized in statistics.

Experimentally, it has been found that the IUPAC approach, based exclusively on the blank variability, in most cases, gives lower values of L_D than the propagation of error approach, which, besides the errors of the blank, takes into account errors in analyte measurement (slope and intersect). Consequently, the propagation of errors approach gives More realistic values of L_D and consistent with the reliability of the blank measures and the signal measures of the standards. In the literature, the propagation of errors is preferred in many chemistry fields [25].

In order to calculate the limit of detection with the propagation of errors approach, it is necessary to make a minimum of five calibration curves to be able to measure s_i and s_m properly, All of these calibration curves would have to be prepared by fortifying control samples with the analyte of interest at concentrations around an estimated limit of detection. This would make the procedure cumbersome for dynamic systems such as chromatography [23].

2.2.4. Root mean square error approach

In this approach, the root mean square error (RMSE) is used instead of the standard deviation of the blank σ_0 in equations 7, 9 and 15, corresponding to L_C, L_D and L_Q, respectively. In order

http://dx.doi.org/10.5772/57341

to calculate the LOD, it is necessary to generate a calibration curve, from which the values of the slope (m) and intersect (i) are obtained. From these values and the equation of the calibration curve a predicted response is calculated (y_p), and then the error associated with each measurement:

$$\left[y_p\text{-}y\right] \tag{18}$$

Then the sum of the square of the errors is calculated for all the points of the calibration curve, and finally the RMSE.

$$RMSE = \left[\frac{\sum_{i=1}^{n}(Y_{ip} - \bar{Y})^2}{n-2}\right]^{1/2} \tag{19}$$

Since the RMSE is calculated from a calibration curve, this approach uses both the variability of the blank and of the measurements. For dynamic systems, such as chromatography with autointegration systems, RMSE is easier to measure and more reliable than σ_0 [23].

2.2.5. The $t_{99sLLMV}$ approach

This method to calculate the MDL analyzed seven fortified samples with amounts of analyte close to an estimated limit of quantitation (ELOQ) or the lowest limit of method validation (LLMV) and its standard deviation $_{sELOQ/LLMV}$ was calculated. This value in a way is a substitute of σ_D in Currie's definitions. Because this approach was developed by EPA, it is used to determine the MDL with the relation:

$$MDL = t_{99n\text{-}1}\, s_{ELOQ/LLMV} \tag{20}$$

Where, $t_{99n\text{-}1}$ is the one sided Student's t for N-1 observations (six degrees of freedom in our case) at the 99% confidence level. In this case $t_{99n=1}$ equals to 3.143.

However, it is extremely important that the ELOQ be accurately determined, because the fortification concentration greatly influenced the final value of MDL and MQL determined by this approach. EPA recommends that if the calculated MQL is significantly different from the ELOQ, the procedure has to be repeated with the calculated MQL as the new ELOQ, and MDL and MQL should be recalculated. This should be done until the calculated values are in the range of the estimated values. This approach is considered a fairly accurate way to determining method detection limits [23]. This approach is similar in some aspects to the so-called empirical method [24, 33], where increasingly lower concentrations of the analyte are analyzed until the measurement do not satisfy a predetermined criteria.

2.2.6. Baseline noise approach

The IUPAC and propagation of errors approaches were developed for spectroscopic analysis. Nearly all concepts used in this approach have an equivalent in chromatography, except the interpretation and measurement of $S_{0,}$ It has been proposed that the chromatographic baseline is analogous to a blank and S_0 must represent a measure of the baseline fluctuations [21, 23].

Therefore, in order to calculate the LOD and LOQ it is necessary to measure the peak-to-peak noise ($N_{p\text{-}p}$) of the baseline around the analyte retention time. $N_{p\text{-}p}$ can be related to the standard deviation of the blank through the relation [21]:

$$s_0 = \frac{N_{p-p}}{5} \tag{21}$$

In spite of being the simplest path to determine the detection capabilities of a chromatographic method, this approach is not recommended because it is very dependent on analyst interpretation since, there is no agreement on where to measure the noise and the extension of baseline that has to be measured. Therefore, the obtained results show great variability between laboratories and even between analysts and consequently, they are hard to compare.

3. Conclusions

The limit of detection is an important figure of merit in analytical chemistry. It is of the utmost importance, in the development of methods to test the detection capabilities of a method and although it is not necessary to calculate it in the process of validation of all methods. It finds applications in areas such as environmental analysis, food analysis and areas under great scrutiny such as forensic science, etc.

Although the detection limit concept is deceptively simple, little is understood by the chemistry community. This caused the proliferation of terms relating to the detection capabilities of a method with different approaches for its determination and impeded efforts to harmonize the methodology.

Although various authors and agencies [20-28, 30-32] have published their own definitions of the detection limit of analytical method, nowadays, the limits of detection and quantification are commonly accepted as that in the hypothesis testing detection limit theory [20, 26, 28, 30-32].

This theory states:

- Limits of detection are actual true values, which can be determined.

- Both Limits are chemical measurement process (CMP) performance characteristics, and therefore, involve all the phases of the analysis. Consequently should not be confused with terms referred exclusively to the detection capabilities of the instrument like IDL.
- Detection limits are not associated with any particular outcome, they are a priori limit
- The existence of both type I errors (false positives) and type II errors (false negatives).
- Detection decision is based on the other hand on a posteriori limit, the critical value.
- Detection limits should not be confused with sensibility, which is the slope of the calibration curve.

In developing the limit of detection theory, Currie made a series of assumptions. First, the measurement distribution of the blanks follow a normal distribution, which is questionable at low concentrations, and secondly, in order to obtain simplified relations, the standard deviation is constant over the range of concentrations studied (homoscedasticity). In this specific circumstances the detection capabilities of a method depends exclusively on blank variability [20, 26, 30-31].

Even Hubaux-Vos prediction bands with their non-uniform width proves that the assumption of homoscedasticity is false. Currie and other authors [26, 28, 30-31] have addressed this problem, but stated that if the standard deviation increases too sharply, limits of detection may not be attainable for the CMP in question.

Although in Currie´s simplified relations, limits of detection exclusively depend on blank variability; other sources of error can be introduced in making the transition from the signal domain to the concentration domain through the uncertainty of the slope and intercept of the calibration curve. To take this into account several approaches have been developed like the propagation error approach, Hubaix –Vos, RMSE, etc [23-26, 32]. Actually, the IUPAC approach which does not account measurement variability, usually gives artificially low values of limit of detection, while methods which account slope and intercept uncertainties, like the propagation error method and Hubaix-Vos method give more realistic estimates, consistent with the reliability of the blank measure and the signal measure of the standards.

In other words, Currie's simplified equations can only be valid, when all their assumptions are met (normal distribution, homoscedasticity, main source of error is the blank). To achieve this, a good knowledge of the blanks is needed to generate confidence in the nature of the blank distribution and some precision in the blank RSD is necessary; therefore an adequate number of full scale true blanks must be assayed through all the CMP.

Most of the assumptions of the IUPAC method are fulfilled in spectrophotometric analysis, for which it was developed, and where it has been used successfully. However, in the case of gas chromatography, where dynamic measures are carried out, and no practical rules are defined to measure the blank standard deviation; the error associated to the intersect of the calibration curve is not always negligible, and the presence of interferences is important. It is better to use a method that takes into account these sources of error. Therefore, for chromatographic techniques it is not recommended the IUPAC approach for the calculation of the detection capabilities of the method. Instead, the propagation of error, Hubaix- Vos, RMSE and $t_{99sLLMV}$ approaches, which take into account the errors of the measurements of the analyte through a calibration curve, are recommended. A brief comparison of the different approaches for the determination of the detection capabilities of a CMP can be found in Table 4.

Method	Easy to apply	Variability of calibration curve	Considers method efficiency and matrix effect	Variability between laboratories and analysts
Ks_0	No	No	Yes	Moderate
$N_{p\text{-}p}$	Yes	No	Yes	High
Propagation of errors	No	Yes	No	Low
Hubaux - Vos	No	Yes	No	Low
RMSE	Yes	Yes	No	Low
$t_{99sLLMV}$	Yes	Yes	Yes	Low

Table 4. Comparison of approaches for calculating detection and quantification limits for analytical method

Since several methods can be used, and could be a difference in the limit of detection calculated by them, it is important that when reporting values of limits of detection, the method used to define these values should be clearly identified in order to have meaningful comparisons.

In order to properly determine the limit of detection and limit of quantification of a method, it is necessary to know the theory behind them, to recognize the scope and limitations of any approach, and be able to choose the method that better suits our CMP. The intention of this chapter is to review the fundamentals of detection limits determination for the purpose of achieving a better understanding of this key element in trace analysis, in particular and analytical chemistry in general; and to achieve a more scientific and less arbitrary use of this figure of merit with a view to their harmonization and avoid the confusion about them, which still prevails in the chemical community.

Nomenclature

α	Type I error, false positive
ACS	American Chemical Society
ADL	Analytical Detection Level
AML	Alternative Minimum Level
API	American Petroleum Institute
ASTM	American Society for Testing and Materials

http://dx.doi.org/10.5772/57341

β	Type II error, false negative
CDER	Conference for Drug Evaluation and Research
CMP	Chemical Measurement Process
CRQL	Contract Required Quantitation Limit
DL	Detection limit/level
DLR	Detection Limit for Purpose of Reporting
$\delta_{\alpha,\beta,\nu}$	the non centrality parameter of the non central t distribution
EDL	Estimated Detection Limit
ELOQ	Estimated limit of quantification
EPA	(United States) Environmental Protection Agency
EQL	Estimated Quantitation Limit
FDA	United States Food and Drug Administration
GLP	Good Laboratory Practice
i	Intersect of the calibration curve
ICH	International Conference for Harmonization
IDL	Instrumental Detection Limit
IML	Interim Minimum level
IQE	Interlaboratory Quantication Estimate
ISO/IEC	International Standards Organization
IUPAC	International Union of Pure and Applied Chemistry
K_α	the one sided tails of the distribution of the blank corresponding to probability levels, 1- ▪ .
K_β	the one sided tails of the distribution of L, when $L=L_D$ corresponding to probability levels, 1- ▪ .
L	Quantity of interest
L_C	Critical value
L_D	Minimum detectable value
L_Q	Minimum Quantification level/value
LLMV	Lower limit of method validation
LOD	Limit of Detection
LOD*	Limit of detectability
LOM	Limit of Measurement
LOQ	Limit of Quantification
m	Sensitivity, slope of calibration curve
MAL	Minimum Analytical Level

MDL	Method Detection Limit
MDL*	Minimum Detection Limit
ML	Minimum Limit
MLD	Minimum Level of Detectability
MQL	Method quantitative limit
MQL*	Minimum quantification level
MQL**	Minimum analytical quantification level
$N_{p\text{-}p}$	Peak to peak noise
NJQL	New Jersey Quantitation Level
PQL	Practical Quantification Level
Pr	Probability
QL	Quantification Level
RMSE	Root mean square error
RSD_Q	Relative standard deviation when $L=L_Q$
S	Analyte signal
σ_0	Standard deviation of the blank
σ_D	Standard deviation when $L=L_D$
σ_Q	Standard deviation when $L=L_Q$
$S_{ELOQ/LLMV}$	Standard deviation when L=ELOQ or L=LLMV
SQL	Sample Quantitation Limit
$t_{1-\alpha,v}$	Student's t, with a probability ▪ ▪ and ▪ degrees of freedom
t_{99n-1}	One sided Student's t, for N-1 observations at the 99% confidence level
USP	United States Pharmacopeia
x	Amount or concentration
x_D	The lowest content it can be distinguished from zero
y_C	The lowest measurable signal
y_p	Predicted response

Acknowledgements

I would like to acknowledge to the Chemical Engineer Sarai Cortes for the exchange of ideas and knowledge on this subject and the joint work.

http://dx.doi.org/10.5772/57341

Author details

Ernesto Bernal[1,2*]

Address all correspondence to: ernestob066@gmail.com

1 Faculty of Criminology, University of Ixtlahuaca CUI, Mexico, Ixtlahuaca, Estado de Mexico, Mexico

2 Chemistry Department, Forensic Science Institute of Mexico City, México

References

[1] Hirsch AF. Good laboratory practice regulations. New York: Marcel Dekker; 1989.

[2] EURACHEM Guide: The fitness for purpose of analytical methods. A Laboratory Guide to method validation and related topics. Teddington: LGC; 1998.

[3] Title 21of the US Code of federal regulations: 21 CFR 211, Current good manufacturing practice for finished pharmaceuticals. Washington: U.S. Government Printing Office; 2012.

[4] Title 21of the US Code of federal regulations: 21 CFR 58, Good laboratory practice for nonclinical laboratory studies. Washington: U.S. Government Printing Office; 2012.

[5] Title 21of the US Code of federal regulations: 21 CFR 320, Bioavailability and bioequivalence requirements. Washington: U.S. Government Printing Office; 2012.

[6] ICH Q7A: Good Manufacturing practice guide for active pharmaceutical ingredients. Rockville: ICH; 2001.

[7] USP 32- NF27. General chapter 1225, Validation of compendial methods. Rockville: USP; 2009.

[8] ISO/IEC 17025, General requirements for the competence of testing and calibration laboratories. 2nd Edition, Switzerland: ISO/IEC; 2005.

[9] ISO 9000: Quality management systems – Fundamentals and vocabulary. Geneva: ISO; 2005

[10] ISO, Guide to the expression of uncertainty in measurement. Geneva: ISO; 1995

[11] Youden WJ, Steiner EH. Statistical manual of the AOAC. Gaithersburg: AOAC International; 1975

[12] Validation of analytical procedures: text and methodology Q2(R1). Rockville: ICH; 2005

[13] Validation of analytical methods for food control. Report of a joint FAO/IAEA expert consultation. Rome: FAO; 1998

[14] Thompson M, Ellison RL, Wood R. Harmonized guidelines for single laboratory validation of methods of analysis. Pure Appl. Chem. 2002; 74(5), 835–855.

[15] U.S. FDA – Guidance for Industry (draft): Analytical Procedures and Methods validation: chemistry, Manufacturing, and Controls and Documentation, Rockville: FDA; 2000

[16] Lopez Garcia P, Buffoni E, Pereira Gomez F, Vilchez Quero JL. Analytical method validation. Wide spectra of quality control. In Akyar I. (ed.), In Tech. Rijeka: In Tech; 2011

[17] Huber L. Validation of analytical method. A primer. Germany: Agilent; 2010

[18] Reviewer Guidance, Validation of Chromatographic Methods. Rockville: Center for Drug Evaluation and Research; 1994

[19] Validation and Peer Review of US EPA Chemical Methods of Analysis, EPA; 2005

[20] Currie LA. Nomenclature and evaluation of analytical methods, including quantification and detection capabilities. Pure Appl. Chem. 1995; 67(10) 1699-1723.

[21] Foley JP, Dorsey JG. Clarification of the limit of detection in chromatography. Chromatographia. 1984; 18 (9) 503-511.

[22] International Organization for Standardization. Capability of detection. Part 1: Terms and definitions. ISO-11843-1.Geneve: ISO; 1997

[23] Corley J. Best practices in establishing detection and quantification limits for pesticide residues in foods, in Lee P. (ed.) Handbook of residue analytical methods for agrochemicals, West Sussex. John Wiley and sons; 2003. p59-75.

[24] Armbruster DA, Tillman MD, Hubbs LA. Limit of detection (LOD)/limit of quantification (LOQ): comparison of the empirical and the statistical methods exemplified with GC-MS assays of abused drugs. Clin. Chem. 1994; 40(7) 1233-1238.

[25] Long GL, Winefordner JD. Limit of detection, a closer look at the IUPAC definition. Anal.Chem. 1983; 55(7), 712A-724A.

[26] Currie LA. Limits for qualitative detection and quantitative determination. Anal.Chem. 1968; 40(3) 586-593.

[27] Chang KH. Limit of detection and its establishment in analytical chemistry. Health Env. J. 2011; 2(1) 38-43.

[28] Lavagnini I, Magno F. A statistical overwiew on univariate calibration, inverse regression and detection limits. Application to GC/MS technique. Mass Spect. Rev. 2007; 26, 1-18.

http://dx.doi.org/10.5772/57341

[29] API, Analytical detection and quantification limits: survey of state and federal approaches. Pub. No. 4721, Washington: American Petroleum Institute; 2002

[30] Currie LA. Detection and quantification limits: origins and historical overview. Anal. Chim. Acta 1999; 391(2) 127-134.

[31] Currie LA. Detection: international update, and some emerging di-lemmas involving calibration, the blank and multiple detection decisions. Chem. Intell. Lab. Sys. 1997; 37, 151-181.

[32] Hubaux, A. Vos, G. Decision and detection limits for linear calibrations curves. Anal. Chem. 1970; 42(8) 849-855.

[33] Needleman SB, Romberg RW. Limits of linearity and detection for some drugs of abuse. J. Anal. Tox. 1990, 14(1) 34-38.

Chapter 4

Gas Chromatography in Metabolomics Study

Yunping Qiu and Deborah Reed

Additional information is available at the end of the chapter

http://dx.doi.org/10.5772/57397

1. Introduction

The metabolome of a cell, tissue, organ, or organism is represented by its small molecular metabolites (molecular weight less than 1000 Da), which are the end products of cellular processes. Variation of the metabolome reflects the interaction of changes in upstream molecules, such as genes and proteins with environmental factors. Metabolic profiling, which is the high-throughput characterization of the metabolome, can be used to assess health status and is a potential diagnostic tool for human diseases. Historically, the sweetness of urine was used to diagnose diabetes. In late 1940s and early 1950s, Roger Williams and colleagues introduced the concept of a "metabolic pattern" to explain the results from their paper chromatography studies that distinguished saliva and urine components among individuals [1]. However, this concept was not widely accepted until the 1960s, when gas chromatography (GC) was successfully used for profiling a complex biological matrix. In the 1960s and 1970s, GC and gas chromatography-mass spectrometry (GC-MS) were successfully used to diagnose metabolic disorders, including maple syrup urine disease [2] and phenylketonuria [3]. In the early 1970s, the term "metabolic profiles" was conceived by Hornings to describe the chromatographic pattern associated with bio-fluid analysis [4].

With the improvement of the sensitivity for the instrument and statistical analysis tools, the field of metabolomics (metabonomics) consequently developed at the end of last century and has been widely used in the past decade. In addition to GC-MS, nuclear magnetic resonance (NMR) and other chromatography coupled mass spectrometry, such as liquid chromatography-mass spectrometry (LC-MS) and capillary electrophoresis–mass spectrometry (CE-MS), are widely used as analytical tools for metabolomics. Compared with other analytical tools, GC-MS is one of the most efficient, sensitive, and reliable tools for metabolomics studies. GC-MS produces reproducible molecular fragmentation patterns making it an integral tool for metabolite identification. Although the application of GC is limited to volatile compounds (before or after derivatization), a large portion of small molecular metabolites are within the range of GC separation. For their different metabolic windows among NMR, LC-MS, CE-MS, and GC-MS, the combination of two or more analytical platforms has been used in many metabolomics studies.

In this chapter we introduce GC-MS-based metabolomics. We will begin with the introduction of GC-MS-based metabolomics procedures, and then describe each section of the procedures in more detail. GC-MS-based metabolomics have been utilized in various areas (such as biomedical research, plant research, and microbial research) as well as for different types of samples. We will also provide some examples of using GC-MS-based metabolomics studies in these described fields.

2. Overview of GC-MS-based metabolomics

Prior to GC-MS analysis, effective sample collection and processing methods are necessary to optimize analyte yield and instrument performance. The collected samples can be liquid, solid, or gas, and different sample pretreatment procedures can be used on different sample types. For complex matrices and solid samples such as blood, tissue, and cell pellets, metabolite extraction is necessary to diminish the matrix effect arising from the complex biological materials. Unlike LC-MS and NMR-based metabolomics, compound derivatization is necessary to increase the volatility of metabolites containing polar functional groups, such as carboxylic and amino groups. The volatile compounds (pre- or post-derivatization) are subjected to GC-MS analysis and data collection. To understand the complex mass signals, a data processing procedure should be performed to identify the true signals, assign the signals into different compounds, and align these compounds from different samples. The aligned data can then reveal the differential variables with different sample groups based on statistical analysis. After peak annotation, the metabolic pathway associated with certain physiological or pathological variations can be obtained. Therefore, the entire process is comprised of sample collection, metabolites extraction, compound derivatization, instrument analysis, data analysis, and metabolite annotation and pathway analysis (Figure 1).

3. Sample collection

Typical samples for GC-MS-based metabolomics can be bio-fluids, tissue, or cell samples. For biomedical or clinical use, bio-fluids are widely used due to the non-invasive collection procedures. Bio-fluids have also been considered to be good sources of diagnostic biomarkers. The most widely used bio-fluids are serum/ plasma and urine. Saliva, tears, and sweat are also used in metabolomics studies. Tissue samples can be obtained from animals, human beings, or plants; and cell samples can be from mammals, plants, or microorganisms. Specific sample collection procedures may impact reproducibility, sensitivity and yield of identified metabolites. Here, we will use blood sample collection and treatment as an example to show how the procedure can impact the metabolomics results.

http://dx.doi.org/10.5772/57397

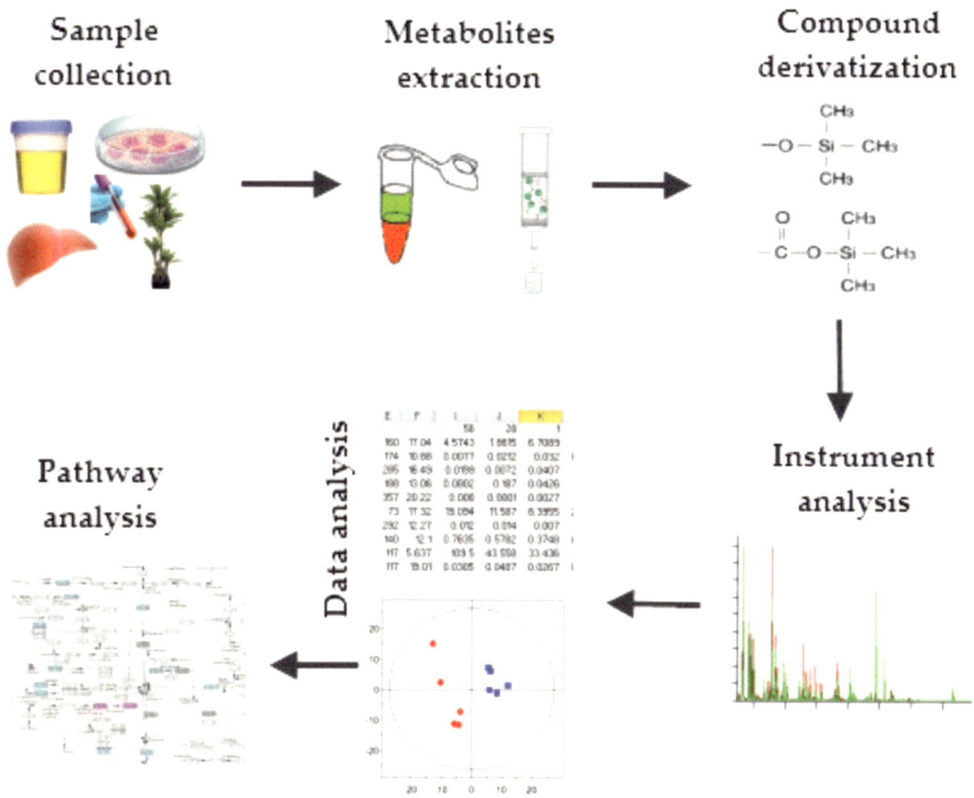

Figure 1. The procedure for GC-MS-based metabolomics.

3.1. Serum or plasma

For blood samples, both serum and plasma can be obtained using different biochemical processes after collection. Serum samples are subjected to a clotting process, which mainly removes fibrinogens, while plasma samples need an anticoagulant during the separation from blood cells. Do these different processes cause variation in metabolites? Is serum or plasma better for metabolomic studies? To answer these questions, previous studies were performed using different analytical platforms. The results showed that there were only slight differences between plasma and serum samples. Using NMR, Teahan et al. showed minor peak shifts between serum and plasma (heparin used as the anticoagulant) samples, such as intensity differences in the resonance dominated by lipid/ triglyceride peaks [5]. In an LC-MS-based untargeted metabolomics study, differences between the serum and plasma (heparin used as the anticoagulant) samples were found to be peptide/protein fragments and lysophophatidylinositol. A lower concentration of lysophophatidylinositol in serum samples was due to degradation in the blood clotting cascade [6]. In an LC-MS/MS-based targeted metabolomics study [7], the authors compared 122 metabolites (focus on lipids and amino acids) between serum and plasma (EDTA used as the anticoagulant) samples. That study found a high correlation between serum and plasma concentrations in all of the targeted metabolites and a slightly higher concentration of metabolites in serum samples. There was also better reproducibility with the plasma samples, but higher sensitivity with the serum samples.

In addition to NMR and LC-MS, GC-MS-based metabolomics has also been used to evaluate the differences between serum and plasma samples. Dettmer et al. showed slightly higher concentrations in most of the 26 quantified metabolites (19 amino acids, 6 organic acids, and glucose), with the exception of pyruvate, malate, citrate, and glucose in serum samples compared to plasma samples (EDTA used as the anticoagulant) [8]. Only lactate and citrate had variation greater than 15%, which could be attributed to serum and plasma sample preparation. The clotting time for serum preparation may result in a higher level of lactate in the serum due to glycolysis occurring in the blood cells. The conjugation of cations (such as Ca^{2+} and Mg^{2+}) with EDTA (the anticoagulant) lowers the concentration of cations in the plasma, which conjugate with citrate during the metabolite extraction process [8]. Slightly higher metabolite concentrations in the serum samples were observed using different analytical platforms, which may be due to lower protein levels in the serum samples as a result of fibrinogen removal during the clotting procedure. Taken together, data from multiple analytical platforms indicate that serum and plasma samples generate similar metabolomics results.

3.2. Pre-analytical variations in serum and plasma samples

In both serum and plasma samples, metabolite variations can be derived during the sample collection and preparation procedures. Therefore in order to acquire high quality samples for metabolomics studies, a suitable protocol for sample collection is very important. In a NMR-based metabolomics study, Teahan et al. analyzed metabolite variations under different processing conditions, and found slight differences in serum samples processed with different clotting times (up to 180 min) regardless of the temperature used (ice or room temperature) [5]. NMR spectrometry has also detected variations between samples caused by freeze-thaw cycles, and therefore it is important to minimize the number of freeze-thaw cycles during sample handling. Serum samples were found to be stable for 24 hours at 4^0C and 4 hours at room temperature [9].

4. Metabolite extraction

Due to the complexity of metabolites with respect to their chemical structures and properties, molecular weight, and concentrations, it is impossible to extract all metabolites using a single extraction procedure. Different extraction procedures target different metabolites. Although many extraction procedures are currently used, such as liquid-liquid extraction, solid-phase extraction, and supercritical fluid extraction, liquid-liquid extraction is used more universally for metabolomic studies. For polar metabolites, commonly used extraction solvents include water, isopropanol, methanol, and acetonitrile, while for lipophilic metabolites, chloroform and ethyl acetate are commonly used. To increase the extracted metabolites range, combinations of extraction solvents, such as methanol/ chloroform or methanol/ chloroform/ water, are used in many metabolite extraction procedures.

Methanol or methanol/chloroform combination is widely used for GC-MS-based metabolomics studies of blood samples (serum/ plasma). Using the combination of a D-optimal experimental design and partial least squares projections to latent structures analysis, Jiye and colleagues developed a methanol/water-based metabolites extraction procedure from plasma [10]. In this study, the authors compared different solvents and their combinations including water, methanol, ethanol, acetonitrile, acetone, and chloroform. Using methoxyamine and N-Methyl-N-trimethylsilyltrifluoroacetamide (MSTFA) derivatization, the authors resolved 501 peaks in a gas chromatography-Time-of-Flight mass spectrometry (GC-TOF-MS) analysis and identified 80 metabolites from these peaks. Their study showed 100 μL of plasma mixed with 900 μL of methanol and water (8:1, v/v) provided optimal metabolites extraction results. Our laboratory has found that the combination of methanol and chloroform (3:1, v/v) is better for serum metabolite extraction than methanol alone [11]. Alternative ratios of methanol and chloroform have also been used in other metabolomics studies. For example, Nishiumi and colleagues used 250 μl of a solvent mixture (MeOH: H_2O: $CHCl_3$ = 2.5:1:1) for extracting 50 μl of serum [12].

Liquid extraction is necessary for metabolomics studies with solid samples, such as tissue and cell pellets. Tissue samples (from mammalian or plant) stored in liquid nitrogen need to be ground to a fine powder prior to solvent extraction or homogenized in a solvent with blenders. A combination of chloroform/methanol/water has been used extensively in metabolite extraction from tissue. Denkert et al. used a combination of chloroform/methanol/water (2:5:2, v/v/v) for colon tissue extraction [13]. Pan et al. developed a two-step metabolite extraction procedure with a mixed solvent of chloroform/methanol/water (1:2:1, v/v/v) in the first step and methanol in the second step for GC-TOFMS-based metabonomics study of liver tissue [14]. In addition, Gullberg et al. optimized an extraction protocol that yielded 66 metabolites from *Arabidopsis thaliana* samples. The authors found that the optimal extraction mixture for plant tissue was 200 μL of chloroform in a vibration mill followed by 800 μl of a methanol/water mixture (75:25, v/v) to form a monophase [15].

The extraction procedures for collected cell pellets have been widely tested with different solvents and various types of cells. Yeast is one of the most investigated species used for targeted metabolites analysis. One commonly used intracellular procedure for metabolite extraction is based on boiling ethanol, which was firstly described by Entian et al. in 1977 [16]. However, boiling ethanol may be harmful for some metabolites. Villas-Boas et al. tested 6 extraction conditions for yeast, including chloroform/methanol/buffer, boiling ethanol, perchloric acid, potassium hydroxide, methanol/water, and pure methanol. Their results indicated that pure methanol is the most suitable solvent and perchloric acid is the least suitable for metabolite extraction [17]. After testing five extraction protocols, including methanol, methanol/chloroform, perchloric acid, boiling ethanol and potassium hydroxide, Winder et al. also found that cold methanol with repeated freeze/thaw cycles was most effective for extracting intracellular metabolites from *E. coli.* [18]. For mammalian cells, methanol, methanol/water, and methanol/chloroform have been identified as suitable agents for metabolite extraction. Methanol/chloroform and boiling methanol with tricine buffer have been shown to sufficiently extract metabolites involved in cellular energy metabolism from canine kidney cells [19]. To increase the range of extracted metabolites, Sellick et al. reported a procedure using two extractions in Chinese hamster ovary cells: the first with pure ethanol and the second with water [20]. In addition, Dettmer et al. found that a methanol/water extraction yielded the best results compared to pure methanol, pure acetone, acetone/water, methanol/chloroform/water, methanol/isopropanol/water, or acid–base methanol in a colorectal cancer cell line (SW480). They also found that direct scraping with methanol/water extraction markedly improved the concentration of metabolites compared to trypsin/EDTA detachment [21].

5. Derivatization

GC-MS is designed to analyze volatile compounds that can pass through a gas chromatography column heated to approximately 300°C. Most metabolites have polar functional groups, such as hydroxyl, amino, or carboxyl groups, and therefore are not volatile at the highest temperature allowed by the GC system. Although some of the metabolites may pass through the GC system, their peak shapes and/or recoveries are often unacceptable due to column absorption. Therefore, protection of those polar functional groups from metabolites with chemical derivatization is necessary.

To analyze as many metabolites as possible, wide range silylation is still the most popular derivatization procedure prior to GC-MS metabolomics analysis. Silylation agents can react with nearly all polar functional groups, including –COOH, –OH, –NH, and –SH. In general, the silylation reaction and stability of the derivatized metabolites decrease in the following order: –hydroxy > hydroxyl (phenol) > carboxylic acid > amine > amide; and primary > secondary > tertiary for alcohols and amines. After silylation, metabolites are more volatile and stable to high temperature. The most commonly used silylation agents for metabolomics include N,O-bis-(trimethylsilyl)-trifluoroacetamide (BSTFA), MSTFA, and N-methyl-N-tert-butyldimethylsilyltrifluoroacetamide (MTBSTFA) (Fig. 2). Typically, 1% trimethylchlorosilane (TMCS) is added as a catalyst for the reaction. Fiehn et al. recommended MSTFA instead of BSTFA for metabolomics studies to avoid high volatility and volatile by-products, including trifluoroacetamide, which causes some interference with early eluting peaks [22]. For keto (-oxo) groups, direct derivatization with silylation will result in multiple products via enolization reactions, which can result in complicated chromatograms with quantification and identification problems. Methoximation is always used to protect keto groups prior to silylation in metabolomic studies, although it will generate two peaks with *syn* and *anti* isoforms for the same keto-containing metabolite. For better separation between the *syn* and *anti* isoforms, Fiehn et al. recommended alkoxyamines instead of hydroxylamines [22].

Since silylation agents are sensitive to moisture, it is important to dry the samples thoroughly prior to the derivatization and prevent moisture exposure during the derivatization procedure. The methoximation reaction can be completed in 1-2 hours with heat or overnight at room temperature [10, 15]. Silylation reactions are usually finished in an hour at room temperature or with heat at 60-70 ^{0}C.

An alternative to TMS silylation is chloroformate derivatization, which has also been used in metabolomic studies. Compare to silylation, the reaction with chloroformates can be conducted in aqueous media, which is very convenient for bio-fluids such urine, serum, and saliva. Reaction in the aqueous phase allows derivatization and analysis of certain volatile polar metabolites, such as short chain fatty acids [23], which would evaporate from the sample during the drying procedure prior to TMS derivatization. In addition, the reaction with chloroformates is very quick; under ultra-sonication, the reaction can be finished in seconds [24]. Chloroformate derivatization is widely used in quantitative measurements of amino acids, organic acids, and amines prior to GC-MS analysis [24]. Qiu et al. successfully adapted an ethyl chloroformate derivatization procedure for urine metabolomics analysis, which was validated with 25 standards (amino acids, amines, and organic acids) and urine samples from a precancerous rat model [25]. The authors used a two-step derivatization to increase the metabolite range and intensity, and found the method to be efficient and robust. Tao et al. subsequently optimized an ethyl chloroformate derivatization procedure for a serum metabolomics study and successfully discriminated between uremic patients and normal subjects [26]. However, only 50 metabolites were identified in that analysis, and thus this method has not been widely used in serum metabolomic studies.

Chloroformate derivatization-based metabolomic analysis has also been used for microbial cells. As reported by Smart et al., a methyl chloroformate (MCF) derivatization-based GC-MS metabolomics analysis of microbial cells was able to profile the end products and/or intermediates from specific metabolic pathways, including glycolysis, the Kreb cycle, amino acid metabolism, and fatty acid metabolism [27]. The authors described the advantages of MCF derivatization as being an economical reagent with a fast derivatization procedure (1 min) and less damage of the GC capillary column by the extraction solvent (chloroform).

Silylation and chloroformate derivatization are the two predominantly used procedures for GC-MS-based metabolomic studies. Figure 2 illustrates examples of some of the derivatization agents react with glycine.

6. Data analysis in GC-MS-based metabolomics

Data analysis for metabolomics study includes two key steps. The first step is to acquire a readable dataset from mass spectrometry signals. The other step is to perform statistical analysis, which allows comparisons of the same data generated from different analytical platforms. However, different types of instruments generate different signal patterns. Therefore, in order to obtain better interpretation of the signals, specific computational programs should be designed.

6.1. Signal interpretation for GC-MS-based metabolomics

The field of metabolomics studies the entire metabolome in a given biological sample. Compared to conventional quantitative analytical studies, which target a limited number of compounds, interpretation of mass signals from metabolomics is substantially more complex

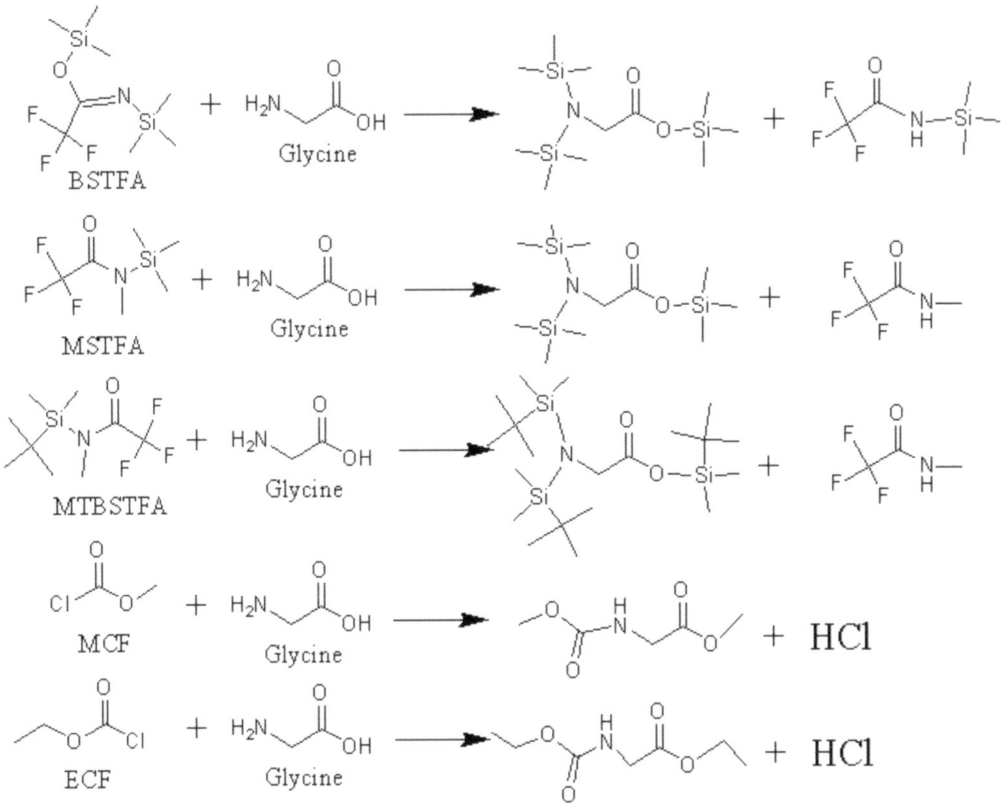

Figure 2. Silylation and chloroformates derivatization with glycine.

and difficult. This complexity lies in two main aspects: deconvolution of coeluting metabolites and alignment of metabolites across multiple samples. An automated mass spectral deconvolution and identification system (AMDIS) developed by Stephen E. Stein is regarded as one of the most powerful software programs for peak deconvolution [28]. This software is designed to deconvolute coeluting compounds from a single sample and does not provide alignment across samples. More recently, several software programs have been designed for GC-MS-based metabolomic data analysis.

Hierarchical Multivariate Curve Resolution (H-MCR) is one of the pioneering software programs that focuses on GC-MS-based metabolomics data [29]. Unlike conventional procedures, this software used a multiprocessing approach to process all interested samples simultaneously. The generated results include the aligned peak information for all samples. However, it will slow down the speed or even exceed computer's capacity when simultaneously processing a large size of samples. An improved procedure was subsequently developed that allows the software to process a subset of representative training samples and generate H-MCR parameters. Using the same parameters, the remaining samples can then be processed very quickly [30].

In 2008, Kopka and colleagues from the Max Planck Institute published a data processing procedure for GC-MS-based metabolomics called TagFinder [31], which is based on the Java programming language. In their process, the raw chromatography files with interchanged format (such as the NetCDF format) or pre-processed peak lists are acceptable. Based on the retention index or retention time, the fragment ions from different chromatograms are binned to "mass tags". After testing of the time groups, the selective mass tags can then be extracted. This process supports both non-targeted fingerprinting and targeting profiling, and the software is freely available for academic use from http://www-en.mpimp-golm.mpg.de/03-research/researchGroups/01-dept1/Root_Metabolism/smp/TagFinder/index.html.

The Automated Data Analysis Pipeline (ADAP; available at http://www.du-lab.org) was developed by Wenxin et al. in 2010. ADAP uses data generated by GC-TOFMS and contains features such as peak detection, deconvolution, peak alignment, and library search [32]. This software is written in standard C++ and R language. To increase the speed of the peak detection and deconvolution steps, parallel computing via Message Passing Interface (MPI) was used. The software was subsequently modified with improved algorithms mainly for peak deconvolution of coeluting metabolites, and substantial improvements in ADAP-GC 2.0 compared to ADAP 1.0 have been reported [33].

In addition to software designed for GC-MS-based metabolomic studies, some software originally developed for LC-MS data has also been used for GC-MS data analysis. XCMS is one of the most popular software programs for MS-based metabolomic studies, and it was originally developed based on LC-MS data [34]. This software was built using an R environment. After data input, a filter and peak identification are performed. The identified peaks are subjected to peak matching across samples. In this step, a group of metabolites are used as standards for a nonlinear retention time correction. Missing values during the peak matching step are filled in during the following step. Statistical analysis and peak visualization are the last two steps. XCMS is freely available under an open-source license at http://metlin.scripps.edu/download/. An online version includes all the same features and is available at https://xcmsonline.scripps.edu [35]. Based on the results from XCMS, a meta-comparison among different groups or complex subgroups can be subsequently performed with metaXCMS (http://metlin.scripps.edu/metaxcms/) [36].

Other LC-MS-based software programs originally designed for LC-MS metabolomics data can also be used for GC-MS-based metabolomic data analysis, including xMSanalyzer [37], MZmine [38], and metAlign [39]. Vendors have also improved their software by adding metabolomic tool kits. For example, ChromaTOF (LECO, St. Joseph, MI) now provides an alignment function with its "Statistical Compare" section.

6.2. Statistical analysis

Metabolomic studies typically contain different groups with multiple samples in each group. The aim of metabolomics is to identify metabolites (variables) that are highly correlated with different conditions, such as gender, stimuli, and pathological status by making comparisons among groups. Due to the complex dataset generated from the chromatography processing, multivariate statistical analysis is typically used in addition to traditional univariate statistics.

Multivariate analyses based on projection methods are popular in metabolomics studies [40, 41]. An unsupervised Principal Component Analysis (PCA) is usually applied to provide a general overview of the acquired dataset and check for outliers. For more specific correlations between the observations and variables, supervised pattern recognitions may be used, such as Partial least-squares (PLS), Orthogonal-PLS (OPLS), PLS-discriminant analysis (PLS-DA), and OPLS-DA. By setting a specific Y matrix, such as gender and pathological status, the variables with high impact on the selected observations can be identified. Those metabolites with high variable importance in the projection (VIP) value can then be selected as differentiating variables. In addition to projection-based methods, other pattern recognition technologies, such as support vector machines (SVM) [42], hierarchical cluster analysis (HCA) [43], and random forest (RF) [44] have also been used in metabolomic data processing.

To further characterize differentiating metabolites, univariate statistics, such as a Student's *t*-test, analysis of variance (ANOVA) test, and non-parametric Mann-Whitney V-tests, are used to compare the selected differential variables from the multivariate statistics [45, 46]. For a large number of distinct variables in the metabolomics data, the false discovery rate (FDR) is used to correct p values from the univariate statistics [47].

7. Metabolite identification and interpretation

Accurate and reproducible compound identification is one of the most important reasons for the popularity of GC-MS in metabolomics studies. However, because of the diversity in chemical structures and concentrations, identification of metabolites remains a challenge. A comparison with commercial libraries, such as the NIST and Wiley libraries, can tentatively identify some of the metabolites. As discussed above, coeluting peaks are a major problem for metabolite identification, and improvements in the separation conditions and advanced deconvolution software (such as ADMIS and ChromaTOF) aid in the accuracy of compound identification. In addition, due to the limited number of metabolite entries in current commercial libraries, even well separated metabolites cannot be identified. Many metabolomic laboratories have developed their own metabolite libraries, which are also used to confirm the identifications based on the retention time (or retention index). FiehnLib is one of the major metabolomic databases and uses a fatty acid methyl ester retention index system. Using GC-MS and GC-TOFMS (LECO, St. Joseph, MI), more than 1,000 primary metabolites, including lipids, amino acids, fatty acids, amines, alcohols, sugars, amino-sugars, sugar alcohols, sugar acids, organic phosphates, hydroxyl acids, aromatics, purines, and sterols have been entered into this library [48].

In addition to biomarker discovery, metabolomics also aims to understand the metabolic variations behind physiological or pathological changes. To obtain the entire picture of metabolic changes, the detected metabolic variations are connected in the metabolic pathways. In a pathway analysis, differential metabolites can be linked with each other as well as to the corresponding genes and enzymes. One of the widely used integrated maps for metabolic pathways, called the 22nd (2003) edition of the IUBMB-Sigma-Nicholson Metabolic Pathways

Chart, is available online (http://www.sigmaaldrich.com/img/assets/4202/MetabolicPathways_6_17_04_.pdf). Other online sources for the metabolic pathway analysis include Kyoto Encyclopedia of Genes and Genomes (KEGG, http://www.genome.jp/kegg), BioCyc (http://www.biocyc.org/), Pathway Commons (PC, http://www.pathwaycommons.org/pc/), Small Molecule Pathway Database (SMPDB, http://www.smpdb.ca/), Reactome (http://www.reactome.org/), The Arabidopsis Information Resource (TAIR, http://www.arabidopsis.org), Expert Protein Analysis System (ExPASy, http://web.expasy.org/pathways/), the University of Minnesota Biocatalysis/Biodegradation Database (UM-BBD, http://umbbd.msi.umn.edu/), and others.

8. Applications of GC-MS-based metabolic profiling

Advanced GC-MS-based metabolic profiling techniques, which evolved with the development of GC-MS instrumentation, were originally applied in biomedical research. However, in the past several decades metabolic profiling has been widely used in plant, microorganism, and environment research as well. In this section, we will introduce applications of GC-MS-based metabolomics in these fields.

8.1. GC-MS-based metabolomics used in biomedical research

Variations in the metabolome are associated with both genetic alterations and environmental interactions. Therefore, the field of metabolomics is an effective tool for biomedical research. GC-MS-based metabolomics approaches are extensively used in biomarker discoveries and biological understandings for human diseases, such as cancers, psychosis, and cardiovascular diseases. Samples can be obtained from both patients and animal models.

Colorectal cancer (CRC) is among the most common types of cancer worldwide. GC-MS-based metabolomics approaches have been used with various types of CRC samples, including urine, serum, tissue, and stool. Qiu et al. used GC-MS analysis of urine from CRC patients and a 1,2-dimethylhydrazine (DMH)-induced precancerous colon lesion rat model and their healthy control counterparts [46]. Good separation was observed between CRC patients and healthy controls, pre- and post-operative patients, and between precancerous colon lesion rats and control rats. The authors observed significantly increased tryptophan metabolism as well as disturbed tricarboxylic acid (TCA) cycle and gut microflora metabolism in both the CRC patients and the rat model. In particular, recovery of tryptophan metabolism toward a healthy state was observed in post-surgical samples. A similar study assessed sera metabolomics from CRC patients and found perturbations in glycolysis, arginine and proline metabolism, fatty acid metabolism, and oleamide metabolism [45]. Ong et al. detected enhancement of the purine salvage pathway in CRC tumor tissues compared to adjacent normal epithelia [49]. Higher concentrations of purines and pyrimidines in CRC tissues were also observed by Denkert et al. based on GC-MS analysis [13]. They also observed higher levels of amino acids and lower levels of intermediates of the TCA cycle and lipids in CRC tissues compared to adjacent normal colon mucosa. In another study, GC-MS analysis was used for metabolic profiling of stool samples from 11 CRC patients and 10 healthy controls, and, higher concentrations of amino acids and lower concentrations of fatty acids and ursodeoxycholic acid were observed in stool samples from CRC patients compared to those from healthy adults[50].

Budczies et al. reported a GC-TOFMS-based metabolic profiling approach for comparing breast cancer tissues to normal ones [51]. That study detected changes in several metabolic pathways, including glycolysis, TCA cycle, nucleotide metabolism, and catabolism of amino acids. They also found that the ratio of cytidine-5-monophosphate to pentadecanoic acid was able to discriminate breast cancer tissues from controls with a sensitivity of 94.8% and a specificity of 93.9%. Using GC-MS analysis, Nam et al. detected four metabolic biomarkers (homovanillate, 4-hydroxyphenylacetate, 5-hydroxyindoleacetate, and urea) in the urine of breast cancer patients, which were validated with gene expression data from the NCBI GEO database [52]. GC-MS-based metabolomics approaches have also been used in several other cancer types, including ovarian carcinoma [53], hepatocellular cancer [54], and osteosarcoma [55].

Several GC-MS-based metabolomics studies on diabetes have also been reported in recent years. To identify metabolic signatures for a relatively new type of diabetes, fulminant type 1 diabetes (FT1DM), Lu et al. profiled the serum metabolome in FT1DM patients together with healthy control subjects, type 2 diabetes, classic type 1 diabetes, and diabetic ketoacidosis patients. Their study found that three metabolite markers (homocysteine, 5-oxoproline, and glutamate) had diagnostic potential for diagnosing FT1DM [56]. GC-MS was also used to evaluate metabolic variations of three drugs, including rosiglitazone, metformin, and repaglinide, in type 2 diabetic patients using multivariate statistical analysis [57]. In another study, two-dimensional GC-TOFMS-based metabolomics coupled with data from LC-MS analysis implicated involvement of the gut microbiota in the development of type 1 diabetes [58].

In addition to cancer and diabetes, GC-MS-based metabolomics approaches have been used in other diseases such as cardiovascular disease and psychosis. To identify metabolite biomarkers for heart failure, Dunn et al. used a GC-MS technique to profile serum metabolites from 52 patients with systolic heart failure and 57 controls [59]. Their observations revealed two promising biomarkers for the failing heart: pseudouridine and 2-oxoglutarate. Yang et al. profiled the serum metabolome in schizophrenia patients and healthy controls using GC-TOFMS and found that fatty acid catabolism was up-regulated with elevated fatty acids and beta-hydroxybutyrate levels in schizophrenia patients [60].

8.2. GC-MS-based metabolomics used in plant research

The plant cell metabolome has not been completely described due to the complexity of primary and secondary metabolites in plants. New metabolites are continuously reported with the development of advanced technologies. GC-MS is widely used in plant metabolomics studies with its powerful separation and compound identification abilities. For example, Fiehn et al. identified 15 relatively new metabolites in *Arabidopsis* extracts by calculating the elemental compositions, and some were found to be completely novel in plant metabolism [22].

One focus in plant metabolomics is the elucidation of the correlation between genotypes and metabolomic phenotypes. The effect of gene modification, identification of key genes and enzymes involved in existing metabolic pathways, or discovery of a new metabolic pathway are widely investigated with metabolomics technology. One example of metabolomics in genetically-modified plants was published in 2000 [61]. Fiehn et al. analyzed 326 distinct compounds in *Arabidopsis thaliana* leaf extracts with GC-MS technology and successfully distinguished four *Arabidopsis* genotypes (two homozygous ecotypes and a mutant of each ecotype) from each other. In another study, Weckwerth et al. applied GC-TOFMS-based metabolomics to a potato plant line with suppressed sucrose synthase isoform II expression levels and found metabolic variations in carbohydrate and amino acid metabolism compared to its parental background line [62].

A more specific link between metabolic regulation and gene or protein expression levels was achieved with the combination of metabolomics and transcriptomic analysis. Using a GC–TOFMS-based metabolomics technique, Albinsky et al. screened the metabolome of *Arabidopsis* lines that overexpressed rice full-length (FL) cDNAs (rice FOX Arabidopsis lines) [63]. Combined with the transcriptome data, they found that Os-LBD37/ASL39 is a new transcriptional regulator for nitrogen metabolism in rice. Kusano et al. also revealed the importance of cytosolic glutamine synthetase 1;1 (*OsGS1*) in the metabolic balance of rice plant growth [64]. In addition, Sulpice et al. attempted to link enzyme activities and metabolite levels to biomass in *Arabidopsis* accessions [65]. Although the association was weak, it is a good example of a metabolomics study using enzyme analysis.

Plant metabolism is largely influenced by genes and genetic modifications; however, the environment can also impact changes in metabolism. Metabolomics is an effective tool for investigating plant adaptive responses against environmental changes. One good example of a link between environment changes and genetic modification was shown in a study using metabolomics technology in *Arabidopsis*. Urano et al. analyzed metabolic variations of *Arabidopsis* with a knock-out mutant of *NCED3* gene wide type compared to the wide type ones under dehydration stress using the combination of GC-MS and CE-MS [66]. Their results showed the dependence of the accumulation of amino acids to abscisic acid production, and the independence of the level of raffinose to abscisic acid production under dehydration stress. Metabolomics technology has also been used to reveal how plants adapt to defend against insects. Using GC-MS analysis, Kant et al. detected a significant increase in the emission of volatile terpenoids in spider mite-infected tomato plants [67].

8.3. GC-MS-based metabolomics used in microbial research

Microorganisms are an essential component of both environmental and human intestinal ecosystems, and they exhibit marked similarities with human beings in terms of metabolism. Microbial metabolomic approaches are important tools for understanding the interactions between human and environmental changes as well as cellular functions.

Escherichia coli represent a commonly found microorganism in the human gut and thus have received extensive attention in previous metabolomics studies. In one time course response study, *E. coli* were exposed to five different conditions (cold, heat, oxidative stress, lactose diauxie, and stationary phase) [68]. The GC-MS-based metabolomic results showed similar global impact on cell metabolism under different stresses, and metabolic signatures revealed a consistent decrease in the levels of metabolites related to glycolysis, the pentose phosphate pathway, and the TCA cycle, which were in agreement with the transcriptomic data. In another example, *E. coli* were used to investigate metabolic responses to different biofuel products, including ethanol, butanol, and isobutanol [69]. The authors observed variations in amino acids and osmoprotectants, including isoleucine, valine, glycine, glutamate, and trehalose in *E. coli* cells exposed to biofuel stress. Stress-derived metabolic variations were also investigated with other microorganisms including yeast. Using GC-MS analysis, Aggio et al. investigated the metabolic changes in yeast cells exposed to different frequencies of sonic vibration and silence and found that metabolic perturbations occurred at different sound frequencies [70].

GC-MS-based metabolomics technology can also be used to rapidly detect microorganisms present in the environment or food. For example, Cevallos-Cevallos et al. developed a rapid GC-MS-based procedure that could simultaneously detect 4 microorganisms in ground beef or chicken: *Escherichia coli O157:H7, Salmonella Typhimurium, Salmonella Muenchen,* and *Salmonella Hartford* [71]. In addition, a study by Sue et al. rapidly detected microbial contamination in fermentation processes using a metabolic footprint analysis approach [72].

9. Conclusions

With its powerful separation and identification ability, GC-MS is widely used in metabolomic studies today. The technology is capable of analyzing non-polar metabolites directly and polar metabolites through derivatization. Many GC-MS-based metabolomics protocols have been developed and validated with various applications, and these applications have resulted in new knowledge regarding metabolic networks as well as enhanced systemic understanding of diseases, metabolic regulations, metabolic response to stresses, and cellular functions. Promising diagnostic biomarkers have also been identified in biomedical research studies, and novel microorganism detection procedures have been developed based on microbial metabolomics. Although great achievements have been made with this technology, challenges still remain. For instance, the processing of huge datasets generated from high-throughput metabolomic studies and identification of those unknown peaks are two of the top challenges for GC-MS-based metabolomics. With continuous development and application, GC-MS-based metabolomics will be a powerful tool that benefits to our daily life.

Author details

Yunping Qiu* and Deborah Reed

*Address all correspondence to: qyp29@163.com

Center for Translational Biomedical Research at University of North Carolina at Greensboro, Kannapolis, NC, USA

References

[1] Gates, S.C. and C.C. Sweeley, *Quantitative metabolic profiling based on gas chromatography.* Clin Chem, 1978. 24(10): p. 1663-73.

[2] Melvin Greera, C.M.W., *Diagnosis of branched-chain ketonuria (maple syrup urine disease) by gas chromatography.* Biochemical Medicine, 1967. 1(1): p. 87–91.

[3] Blau, K., H.H. Cameron, and G.K. Summer, *Diagnosis of phenylketonuria by gas chromatography.* Methods Med Res, 1970. 12: p. 100-5.

[4] E. C. Horning, M.G.H., *Human Metabolic Profiles Obtained by GC and GC/MS.* Journal of Chromatographic Science, 1971. 9(3): p. 129-140.

[5] Teahan, O., et al., *Impact of analytical bias in metabonomic studies of human blood serum and plasma.* Anal Chem, 2006. 78(13): p. 4307-18.

[6] Denery, J.R., A.A. Nunes, and T.J. Dickerson, *Characterization of differences between blood sample matrices in untargeted metabolomics.* Anal Chem, 2011. 83(3): p. 1040-7.

[7] Yu, Z., et al., *Differences between human plasma and serum metabolite profiles.* PLoS One, 2011. 6(7): p. e21230.

[8] Dettmer, K., et al., *Comparison of serum versus plasma collection in gas chromatography--mass spectrometry-based metabolomics.* Electrophoresis, 2010. 31(14): p. 2365-73.

[9] Fliniaux, O., et al., *Influence of common preanalytical variations on the metabolic profile of serum samples in biobanks.* J Biomol NMR, 2011. 51(4): p. 457-65.

[10] A, J., et al., *Extraction and GC/MS analysis of the human blood plasma metabolome.* Analytical chemistry, 2005. 77(24): p. 8086-94.

[11] Qiu, Y., *Metabonomics Study on Colorectal Cancer Using Combined Chromatography-mass Spectrometry Strategy* 2008, Shanghai Jiao Tong University: Shanghai. p. 92-95.

[12] Nishiumi, S., et al., *A novel serum metabolomics-based diagnostic approach for colorectal cancer.* PLoS One, 2012. 7(7): p. e40459.

[13] Denkert, C., et al., *Metabolite profiling of human colon carcinoma--deregulation of TCA cycle and amino acid turnover.* Mol Cancer, 2008. 7: p. 72.

[14] Pan, L., et al., *An optimized procedure for metabonomic analysis of rat liver tissue using gas chromatography/time-of-flight mass spectrometry.* J Pharm Biomed Anal, 2010. 52(4): p. 589-96.

[15] Gullberg, J., et al., *Design of experiments: an efficient strategy to identify factors influencing extraction and derivatization of Arabidopsis thaliana samples in metabolomic studies with gas chromatography/mass spectrometry.* Anal Biochem, 2004. 331(2): p. 283-95.

[16] Entian, K.D., F.K. Zimmermann, and I. Scheel, *A partial defect in carbon catabolite repression in mutants of Saccharomyces cerevisiae with reduced hexose phosphyorylation.* Mol Gen Genet, 1977. 156(1): p. 99-105.

[17] Villas-Boas, S.G., et al., *Global metabolite analysis of yeast: evaluation of sample preparation methods.* Yeast, 2005. 22(14): p. 1155-69.

[18] Winder, C.L., et al., *Global metabolic profiling of Escherichia coli cultures: an evaluation of methods for quenching and extraction of intracellular metabolites.* Anal Chem, 2008. 80(8): p. 2939-48.

[19] Ritter, J.B., Y. Genzel, and U. Reichl, *Simultaneous extraction of several metabolites of energy metabolism and related substances in mammalian cells: optimization using experimental design.* Anal Biochem, 2008. 373(2): p. 349-69.

[20] Christopher A. Sellick, D.K., Alexandra S. Croxford, Arfa R. Maqsood, Gill M. Stephens, Royston Goodacre, Alan J. Dickson, *Evaluation of extraction processes for intracellular metabolite profiling of mammalian cells: matching extraction approaches to cell type and metabolite targets.* Metabolomics, 2010. 6: p. 427–438.

[21] Dettmer, K., et al., *Metabolite extraction from adherently growing mammalian cells for metabolomics studies: optimization of harvesting and extraction protocols.* Anal Bioanal Chem, 2011. 399(3): p. 1127-39.

[22] Fiehn, O., et al., *Identification of uncommon plant metabolites based on calculation of elemental compositions using gas chromatography and quadrupole mass spectrometry.* Anal Chem, 2000. 72(15): p. 3573-80.

[23] Zheng, X., et al., *A targeted metabolomic protocol for short-chain fatty acids and branched-chain amino acids.* Metabolomics, 2013. 9(4): p. 818-827.

[24] Husek, P., *Chloroformates in gas chromatography as general purpose derivatizing agents.* J Chromatogr B Biomed Sci Appl, 1998. 717(1-2): p. 57-91.

[25] Qiu, Y., et al., *Application of ethyl chloroformate derivatization for gas chromatography-mass spectrometry based metabonomic profiling.* Anal Chim Acta, 2007. 583(2): p. 277-83.

[26] Tao, X., et al., *GC-MS with ethyl chloroformate derivatization for comprehensive analysis of metabolites in serum and its application to human uremia.* Anal Bioanal Chem, 2008. 391(8): p. 2881-9.

[27] Smart, K.F., et al., *Analytical platform for metabolome analysis of microbial cells using methyl chloroformate derivatization followed by gas chromatography-mass spectrometry.* Nature protocols, 2010. 5(10): p. 1709-29.

[28] Stein, S., *An integrated method for spectrum extraction and compound identification from gas chromatography/mass spectrometry data* J. Am. Soc. Mass Spectrom, 1999. 10(8): p. 770-781.

[29] Jonsson, P., et al., *High-throughput data analysis for detecting and identifying differences between samples in GC/MS-based metabolomic analyses.* Anal Chem, 2005. 77(17): p. 5635-42.

[30] Jonsson, P., et al., *Predictive metabolite profiling applying hierarchical multivariate curve resolution to GC-MS data--a potential tool for multi-parametric diagnosis.* J Proteome Res, 2006. 5(6): p. 1407-14.

[31] Luedemann, A., et al., *TagFinder for the quantitative analysis of gas chromatography--mass spectrometry (GC-MS)-based metabolite profiling experiments.* Bioinformatics, 2008. 24(5): p. 732-7.

[32] Jiang, W., et al., *An automated data analysis pipeline for GC-TOF-MS metabonomics studies.* Journal of proteome research, 2010. 9(11): p. 5974-81.

[33] Ni, Y., et al., *ADAP-GC 2.0: deconvolution of coeluting metabolites from GC/TOF-MS data for metabolomics studies.* Anal Chem, 2012. 84(15): p. 6619-29.

[34] Smith, C.A., et al., *XCMS: processing mass spectrometry data for metabolite profiling using nonlinear peak alignment, matching, and identification.* Anal Chem, 2006. 78(3): p. 779-87.

[35] Tautenhahn, R., et al., *XCMS Online: a web-based platform to process untargeted metabolomic data.* Anal Chem, 2012. 84(11): p. 5035-9.

[36] Tautenhahn, R., et al., *metaXCMS: second-order analysis of untargeted metabolomics data.* Anal Chem, 2011. 83(3): p. 696-700.

[37] Uppal, K., et al., *xMSanalyzer: automated pipeline for improved feature detection and downstream analysis of large-scale, non-targeted metabolomics data.* BMC bioinformatics, 2013. 14: p. 15.

[38] Katajamaa, M., J. Miettinen, and M. Oresic, *MZmine: toolbox for processing and visualization of mass spectrometry based molecular profile data.* Bioinformatics, 2006. 22(5): p. 634-6.

[39] Lommen, A., *MetAlign: interface-driven, versatile metabolomics tool for hyphenated full-scan mass spectrometry data preprocessing.* Anal Chem, 2009. 81(8): p. 3079-86.

[40] Holmes, E. and H. Antti, *Chemometric contributions to the evolution of metabonomics: mathematical solutions to characterising and interpreting complex biological NMR spectra.* Analyst, 2002. 127(12): p. 1549-57.

[41] Trygg, J., E. Holmes, and T. Lundstedt, *Chemometrics in metabonomics.* J Proteome Res, 2007. 6(2): p. 469-79.

[42] Mahadevan, S., et al., *Analysis of metabolomic data using support vector machines.* Anal Chem, 2008. 80(19): p. 7562-70.

[43] Tikunov, Y., et al., *A novel approach for nontargeted data analysis for metabolomics. Large-scale profiling of tomato fruit volatiles.* Plant Physiol, 2005. 139(3): p. 1125-37.

[44] Chen, T., et al., *Random forest in clinical metabolomics for phenotypic discrimination and biomarker selection.* Evid Based Complement Alternat Med, 2013. 2013: p. 298183.

[45] Qiu, Y., et al., *Serum metabolite profiling of human colorectal cancer using GC-TOFMS and UPLC-QTOFMS.* J Proteome Res, 2009. 8(10): p. 4844-50.

[46] Qiu, Y., et al., *Urinary metabonomic study on colorectal cancer.* J Proteome Res, 2010. 9(3): p. 1627-34.

[47] Boudonck, K.J., et al., *Discovery of metabolomics biomarkers for early detection of nephrotoxicity.* Toxicol Pathol, 2009. 37(3): p. 280-92.

[48] Kind, T., et al., *FiehnLib: mass spectral and retention index libraries for metabolomics based on quadrupole and time-of-flight gas chromatography/mass spectrometry.* Anal Chem, 2009. 81(24): p. 10038-48.

[49] Ong, E.S., et al., *Metabolic profiling in colorectal cancer reveals signature metabolic shifts during tumorigenesis.* Mol Cell Proteomics, 2010.

[50] Weir, T.L., et al., *Stool Microbiome and Metabolome Differences between Colorectal Cancer Patients and Healthy Adults.* PLoS One, 2013. 8(8): p. e70803.

[51] Budczies, J., et al., *Remodeling of central metabolism in invasive breast cancer compared to normal breast tissue - a GC-TOFMS based metabolomics study.* BMC Genomics, 2012. 13: p. 334.

[52] Nam, H., et al., *Combining tissue transcriptomics and urine metabolomics for breast cancer biomarker identification.* Bioinformatics, 2009. 25(23): p. 3151-7.

[53] Denkert, C., et al., *Mass spectrometry-based metabolic profiling reveals different metabolite patterns in invasive ovarian carcinomas and ovarian borderline tumors.* Cancer Res, 2006. 66(22): p. 10795-804.

[54] Chen, T., et al., *Serum and urine metabolite profiling reveals potential biomarkers of human hepatocellular carcinoma.* Mol Cell Proteomics, 2011. 10(7): p. M110 004945.

[55] Zhang, Z., et al., *Serum and urinary metabonomic study of human osteosarcoma.* J Proteome Res, 2010. 9(9): p. 4861-8.

[56] Lu, J., et al., *Serum metabolic signatures of fulminant type 1 diabetes.* J Proteome Res, 2012. 11(9): p. 4705-11.

[57] Bao, Y., et al., *Metabonomic variations in the drug-treated type 2 diabetes mellitus patients and healthy volunteers.* J Proteome Res, 2009. 8(4): p. 1623-30.

[58] Oresic, M., et al., *Dysregulation of lipid and amino acid metabolism precedes islet autoimmunity in children who later progress to type 1 diabetes.* J Exp Med, 2008. 205(13): p. 2975-84.

[59] Warwick B. Dunn, D.I.B., Sasalu M. Deepak, Mamta H. Buch, Garry McDowell, Irena Spasic, David I. Ellis, Nicholas Brooks, Douglas B. Kell, Ludwig Neyses, *Serum metab-*

olomics reveals many novel metabolic markers of heart failure, including pseudouridine and 2-oxoglutarate. Metabolomics, 2007. 3(4): p. 413–426.

[60] 60.Yang, J., et al., *Potential metabolite markers of schizophrenia.* Mol Psychiatry, 2013. 18(1): p. 67-78.

[61] Fiehn, O., et al., *Metabolite profiling for plant functional genomics.* Nat Biotechnol, 2000. 18(11): p. 1157-61.

[62] Weckwerth, W., et al., *Differential metabolic networks unravel the effects of silent plant phenotypes.* Proc Natl Acad Sci U S A, 2004. 101(20): p. 7809-14.

[63] Albinsky, D., et al., *Metabolomic screening applied to rice FOX Arabidopsis lines leads to the identification of a gene-changing nitrogen metabolism.* Mol Plant, 2010. 3(1): p. 125-42.

[64] Kusano, M., et al., *Metabolomics data reveal a crucial role of cytosolic glutamine synthetase 1;1 in coordinating metabolic balance in rice.* Plant J, 2011. 66(3): p. 456-66.

[65] Sulpice, R., et al., *Network analysis of enzyme activities and metabolite levels and their relationship to biomass in a large panel of Arabidopsis accessions.* Plant Cell, 2010. 22(8): p. 2872-93.

[66] Urano, K., et al., *Characterization of the ABA-regulated global responses to dehydration in Arabidopsis by metabolomics.* Plant J, 2009. 57(6): p. 1065-78.

[67] Kant, M.R., et al., *Differential timing of spider mite-induced direct and indirect defenses in tomato plants.* Plant Physiol, 2004. 135(1): p. 483-95.

[68] Jozefczuk, S., et al., *Metabolomic and transcriptomic stress response of Escherichia coli.* Mol Syst Biol, 2010. 6: p. 364.

[69] Wang, J., et al., *Global Metabolomic and Network analysis of Escherichia coli Responses to Exogenous Biofuels.* J Proteome Res, 2013.

[70] Raphael Bastos Mereschi Aggio, V.O., Silas Granato Villas-Boas, *Sonic vibration affects the metabolism of yeast cells growing in liquid culture: a metabolomic study.* Metabolomics, 2012. 8(4): p. 670-678.

[71] Cevallos-Cevallos, J.M., M.D. Danyluk, and J.I. Reyes-De-Corcuera, *GC-MS based metabolomics for rapid simultaneous detection of Escherichia coli O157:H7, Salmonella Typhimurium, Salmonella Muenchen, and Salmonella Hartford in ground beef and chicken.* J Food Sci, 2011. 76(4): p. M238-46.

[72] Sue, T., et al., *An exometabolomics approach to monitoring microbial contamination in microalgal fermentation processes by using metabolic footprint analysis.* Appl Environ Microbiol, 2011. 77(21): p. 7605-10.

Chapter 5

The Utilization of Gas Chromatography/Mass Spectrometry in the Profiling of Several Antioxidants in Botanicals

Serban Moldoveanu

Additional information is available at the end of the chapter

http://dx.doi.org/10.5772/57292

1. Introduction

1.1. Introductory information regarding antioxidants in botanicals

Antioxidants are chemicals that inhibit oxidation, and certain antioxidant molecules from fruits and vegetables are thought to alleviate oxidative stress in biological systems. Oxidative stress is a process generated by excessive reactive oxygen species (ROS) in organisms, and it is considered to be involved in a number of illnesses such as cancer, arteriosclerosis, heart diseases, etc. Among these reactive oxygen species are hydroxyl radical $OH^{\bullet}$, superoxide radical $O_2^{\bullet\,\bullet}$ and also hydrogen peroxide H_2O_2. Reactive nitrogen species (RNS) are also present in organisms, although at lower levels. The RNS include nitric oxide $NO^{\bullet}$, nitrogen dioxide NO_2, nitrosyl cation NO^{+}, etc. Reducing agents are also present in aerobic organisms. Among these are ascorbic acid, glutathione, and uric acid, and these molecules maintain a limited level of ROS in the organism. The enhancement of endogenous antioxidant capability of the human body is thought to be achieved by: 1) ingesting exogenous antioxidants either as food or as dietary supplements, 2) inducing the body production of antioxidant enzymes such as catalase, glutathione peroxidase, and superoxide dismutase, also with ingesting certain compounds, 3) inhibiting lipid peroxidation. The use of specific botanicals, either as food or as dietary supplements, has been intensively investigated for potential health benefits (see e.g. [1-5]). The reaction of antioxidants that interact with free radicals on a one-to-one basis takes place through various mechanisms. Among these are the hydrogen atom transfer (HAT), single electron transfer followed by proton transfer (SET or ET-PT), and sequential proton loss electron transfer (SPLET). For example, the HAT mechanism can be described by the following reactions (where $ROO^{\bullet}$ is a free radical and AH an antioxidant):

$$ROO^{\bullet} + AH \rightarrow ROOH + A^{\bullet}$$

$$ROO^{\bullet} + A^{\bullet} \rightarrow ROOA$$

The mechanism by which antioxidant enzymes are stimulated in the human body is less well understood, but specific botanicals with "antioxidant character" are also recommended for this purpose.

Several procedures have been reported in the literature for the characterization of antioxidant properties of a material (typically food or dietary supplement). Among these are parameters such as "oxygen radical absorbance capacity" or ORAC [6-8], "ferric ion reducing antioxidant power" (FRAP) [9], "Folin-Ciocalteu reducing capacity assay" (FCR) [10], etc. The ORAC parameter can be measured by two versions of the same procedure, one indicated as hydrophilic ORAC and the other as lipophilic ORAC [6] and is expressed as μM of Trolox (TE) per g of sample. FRAP values are expressed in μM Fe^{2+} per g of sample [9]. The chemical nature of the antioxidants from different sources can vary considerably. Each compound may have different antioxidant properties, and may be considered useful for specific health benefits. Also, beneficial synergistic effects were reported for specific associations of compounds [11]. For these reasons, the analysis of individual antioxidant molecules including their identification and quantitation is important. Antioxidants from botanicals belong to different classes of molecules. Among these are the following:

1. Monoterpenoid phenols and alcohols such as: thymol, carvacol, menthol.
2. Diterpene phenols, such as: carnosic acid, carnosol, rosmanol.
3. Hydroxycinnamic type compounds such as: caffeic acid, chlorogenic acid, rosmarinic acid, p-coumaric acid, resveratrol, curcumin, eugenol, cinnamaladehyde.
4. Hydroxybenzoic acids and derivatives such as: gallic acid, protocatechuic acid, propyl gallate, tannins.
5. Benzopyrones (2- and 4-) and xanthones such as: scopoletin, coumarin, quercetin, genistein, naringenin, diosmin, rutin, mangiferin.
6. Flavones and their derivatives such as: epicatechin, epigallocatechin, epicatechin gallate, epigallocatechin gallate, gossypin.
7. Dihydrochalcones, such as aspalathin, notophagin.
8. Anthocyanins and anthocyanidins, such as cyanidin, pelargonidin, cyanidin glucosides.
9. Triterpene acids such as ursolic acid, oleanolic acid, betulinic acid.
10. Tocopherols, such as α, β, γ, δ-tocopherols, tocotrienols.
11. Carotenoids, β-carotene, lutein.
12. Ubiquinone, CoQ10.
13. Ascorbic acid, ascorbyl palmitate.

14. Benzodioxoles, such as myristicin, piperine, safrole.

15. Unsaturated lipids.

16. Other compounds, such as gambogic acid, gingerol, ar-turmerone, antioxidant enzymes.

Most antioxidants molecules are relatively large, and in addition, these molecules are frequently polar with groups such as OH and COOH. The high molecular weight and the high polarity of many antioxidant molecules are not conducive to the use of gas chromatography (GC) as the preferred analytical tool. If the molecule is also thermally unstable, such as lutein and carotene, the use of GC is definitely inadequate. For this reason, the analysis of many antioxidant compounds has been performed using high performance liquid chromatography (HPLC) methods [12-24]. However, GC methods can also be used for the identification of antioxidant compounds [25-27]. The use of mass spectrometric detection with its excellent capability for the determination of compound chemical formula makes GC/MS an irreplaceable tool when antioxidant analysis requires compound identification. Although significant progress has been made in using LC/MS (and LC/MS/MS) for compound identification, these techniques still remain more adequate for quantitation and not for qualitative analysis. Various procedures for the GC/MS analysis of certain antioxidants in botanicals are further described in this chapter.

2. Experimental procedures for extending the GC analysis to larger molecules

The GC/MS analysis has considerable advantages compared to other analytical techniques. Besides the simplicity of the procedure, the technique can be used for definite identification of each compound based on its MS spectrum. Also, GC/MS provides separation with excellent resolution of the compounds, and is suitable for quantitation when standards are available. The area counts in the chromatograms can be measured and expressed as normalized area counts reported to the peak area of an internal standard. This type of presentation of results, does not provide quantitative levels for compounds, but allows for the determination of which sample has a higher or a lower level of a given compound. The disadvantage of the technique is caused by the need for volatility and certain thermal stability for the compounds to be analyzed. These restrictions limit the use of GC to larger and non-volatile molecules. However, several procedures are used for extending the capability of gas chromatography for the analysis of these types of molecules. Among these procedures are specific techniques for sample preparation, in particular the derivatization of the analytes. Other procedures include certain GC instrument settings such as the use of hydrogen as carrier gas, selection of appropriate chromatographic column, selection of the type of injection port, and a GC oven gradient with high final temperatures, etc. Derivatization of analytes can be beneficial in a variety of circumstances in GC, such as when the polarity of the analyte is too high and does not elute from the column, when a desired separation is not achievable, when the peak shape of a compound is not good, or when the analyte is not stable in the injection port of the GC. Many antioxidants fit this scenario, and for this reason derivatization is frequently used in GC/MS analysis of antioxidants from botanicals. A variety of chemical reactions are utilized for analytes derivatization. These reactions can be alkylations, arylations, silylation, acylation, additions to carbon-heteroatom multiple bonds, etc. Hydrolysis and formation of smaller molecules (e.g. from lipids) is another type of chemical reaction used as sample preparation step for GC analysis. Of particular interest for the derivatization of many antioxidant molecules is silylation. Many antioxidant molecules contain OH and COOH groups, and these can be easy derivatized using silylation. For this reason, silylation is a preferred technique used for extending the range of analysis by GC/MS of antioxidants. However, in spite of the utility of GC/MS for antioxidant analysis it must be emphasized that it offers only a limited window in the whole range of antioxidant compounds present in botanicals, and heavier molecules may still need to be analyzed using HPLC methods.

Larger molecules, even after derivatization typically require specific conditions for the GC separation, such as temperature gradient up to a relatively high temperature. Modern GC ovens are designed to be able to reach temperatures as high as 400 °C, but the limiting factor regarding the oven temperature is typically the stability of the stationary phase of the column. Depending on the nature of the stationary phase, the columns may be stable up to 360 °C, and only special ones may stand higher temperatures. Such temperatures are necessary in certain instances for the elution of heavier compounds from the chromatographic column. The typical split/splitless injection port, with relatively high temperatures (e.g. around 300 °C) is frequently adequate for the analysis of larger molecules. However, some compounds decompose in the standard split/splitless injection port and "cold on-column" injection is necessary for obtaining acceptable results [28,29].

The identification of the compounds in the chromatogram is typically performed using the library search capability of the GC/MS instrument and a mass spectral library (e.g. NIST8, NIST11, Wiley_9THL, Wiley Registry 10th Ed., etc.). However, the mass spectra of most antioxidant molecules, in particular in silylated form, are not available in standard mass spectral libraries. For this reason, the identification of unknown antioxidant molecules in a natural product material may be difficult. Valuable information can be obtained from separate analysis of standard compounds (if available), or from the comparison of spectra of unknowns with those of expected similar molecules that are available as standards and can be directly analyzed. Special procedures can be used that help with identification of silylated compounds, by using derivatization with deuterated silylating reagents. As an example, the use of d_9-BSTFA [d_9-bis(trimethylsilyl)-trifluoroacetamide] that generates deuterated TMS (trimethylsilyl) derivatives allows the detection of the number of silyl groups in a compound (and implicitly of the number of OH or COOH groups). This can be done by comparing the masses of the same compound when silylated with the deuterated reagent and when silylated with non-deuterated reagent. For each TMS group a difference of 9 a.m.u. is noticed between the two types of silylated compounds. One additional procedure that helps with the identification of an unknown compound is based on high resolution mass spectra. The spectrum with high resolution can be obtained using MS instruments that were initially recommended for spectra in unit resolution, by using specific post acquisition programs and internal calibration (e.g. MassWorks, Cerno Bioscience, Danbury, CT 06810 USA). Such programs allow the determination of the probable empirical molecular formula of unknown compounds.

Common procedures for the quantitation of specific compounds, such as calibration curves using standards can be applied for the quantitation in case the compound derivatization is not strongly influenced by the sample matrix. In some cases, the standard addition technique for quantitation (see e.g. [30]) gives better results as compared to external calibration. In both cases, the unavailability of standards may limit the possibilities for quantitation.

3. Examples of GC/MS analysis of botanicals containing antioxidant molecules

A large number of botanicals contain antioxidant compounds, and their properties are extensively reported in the literature (see e.g. [31,32]). Also numerous studies were dedicated to individual botanical composition and content of antioxidants. Two examples of botanicals studied using GC/MS of directly silylated natural material (e.g. leaves) are further described. The silylation technique starts with 50 mg solid sample which is weighed (with 0.1 mg precision) in GC vials (2 mL screw top vials with screw caps with septa, Agilent, Wilmington, Delaware 19808). The silylation is done to all the compounds containing active hydrogens, such as acids, alcohols, or amines. The result is the formation of various trimethylsilyl (TMS) derivatives. A reagent and a solvent are used for the silylation process, and the procedure does not require a separate extraction step. From various available reagents, it was determined that bis(trimethylsilyl)-trifluoroacetamide (BSTFA) with 1% trimethylchloro-silane (TMCS) gives the best results. The preferred solvent was found to be N,N-dimethylformamide (DMF). The solvent used in this study contained as internal standard *tert*-butylhydroquinone. The DMF solution with internal standards was prepared using 100 mL DMF and 40 mg *tert*- butylhydroquinone (all compounds from Aldich/Sigma Saint Louis, MO 63178-9916). The final DMF solution contained 0.4 mg/mL *tert*-butylhydroquinone. For the analysis, to each vial were added 0.4 mL DMF with internal standards and 0.8 mL BSTFA with 1% TMCS (Aldrich/Sigma Saint Louis, MO 63178-9916). The vials were kept at 78° C (in a heating block) for 30 min., and were allowed to cool at room temperature for another 30 min. After cooling the solution, each vial was filtered through a 0.45 μm PTFE filter (VWR Suwanee, GA 30024) into screw top vials with screw caps and septa, which were used for the GC/MS analysis. This procedure can be scaled down when either the amount of sample is small, or when using the derivatization reagent d_9-BSTFA (available from CDN Isotopes, Pointe Claire, Canada H9R 1H1). The analysis was done using a GC/MS instrument (such as Agilent 6890/5973 system from Agilent, Wilmington, Delaware 19808). The GC/MS conditions are given in Table 1.

Parameter	Description	Parameter	Description
GC column	DB-5MS*	Carrier gas	Hydrogen
Column dimensions	30 m long, 0.25 mm id.	Flow mode	Constant flow
Film thickness	0.25 μm	Flow rate	1.0 mL/min
Initial oven temp.	50°C	Nominal initial pressure	7.57 psi
Initial time	0.5 min	Split ratio	30 : 1
Oven ramp rate	3°C/min	Split flow	29.8 mL/min
Oven final first ramp	200°C	GC outlet	MSD
Final time first ramp	0 min	Outlet pressure	Vacuum
Oven ramp rate	4°C/min	Transfer line heater	300°C
Oven final temp.	320°C	Ion source temp.	230°C
Final time	10 min	Quadrupole temp.	150°C
Total run time	90.5 min	MSD EM offset	100 V
Inlet temp.	300°C	MSD solvent delay	7.0 min
Inlet mode	Split	MSD acquisition mode	scan
Injection volume	1.0 μL	Mass range	33 to 1050 a.m.u.

*Note: equivalent columns can be used, such as ZB-5 HT Inferno, etc.

Table 1. GC/MS operating parameters for silylated compounds analysis.

As shown in Table 1, the separation used hydrogen as a carrier gas, and a relatively high final oven temperature.

3.1. Example of green tea analysis

Green tea (leaves of *Camellia sinensis*) is a well known botanical with antioxidant properties [33-35]. For the green tea evaluated in this study (commercially available from Shanghai Tiantan Intern. Trading Co., Ltd.) the ORAC values (both lipophilic and hydrophilic) were 1150 ± 20 μM TE/g, and FRAP was 2200 ± 15 μM Fe^{2+}/g. The chemical composition of green tea leaf can be studied using the GC/MS technique after direct silylation of the dry leaf. The chromatogram of silylated green tea is shown in Figure 1.

The identification of the main peaks from Figure 1 can be viewed in Table 2 where the retention times for individual compounds are listed.

Some of the spectra of the silylated compounds are not available in common mass spectral libraries. The spectra of silylated epigallocatechin (EGC), epicatechin gallate (ECG), epigallocatechin gallate(EGCG), and chlorogenic acid, as obtained using standards, are shown in Figures 2 to 5.

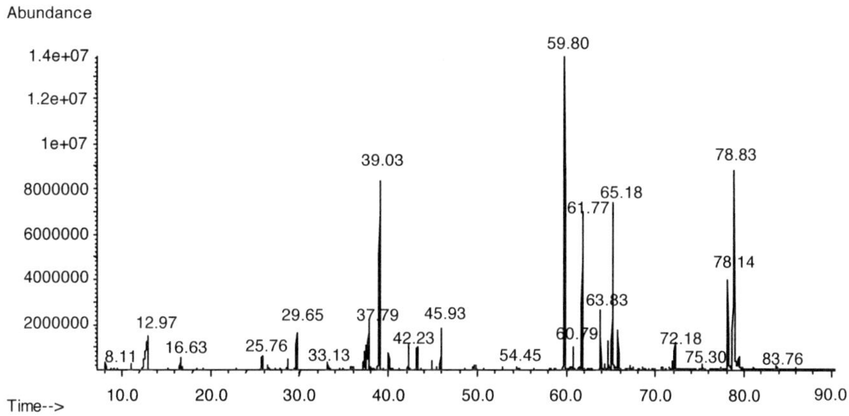

Figure 1. Chromatogram of silylated green tea dry leaf. The identification of main peaks is given in Table 2. (I.S. elutes at 29.65 min).

Antioxidant compounds	Ret. Time	Other main compounds	Ret. Time
Caffeine	37.79	Phosphate	16.63
Gallic acid	42.23	Malic acid	25.76
Epicatechin	63.83	5-Oxoproline	26.42
Catechin	64.32	Fructose	37.20
Epigallocatechin (EGC)	65.18	Quinic acid	39.03
α-Tocoferol (trace)	68.03	Glucose	43.23
Chlorogenic acid (trace)	68.11	Myoinositol	45.93
Epicatechin gallate	78.14	Sucrose	59.80
Epigallocatechin gallate (EGCG)	78.83	Disaccharide	61.77
Gallocatechin gallate	79.41	Disaccharide ?	72.18

Table 2. Compound identification for green tea chromatogram shown in Figure 1

Fragmentation indicated in Figures 2 to 5 can be verified using silylation with d_9-BSTFA. Figure 6 shows the spectrum of d_9-silylated epigallocatechin gallate.

The masses of different ions are explained in Figure 6 in comparison with those shown in Figure 5. For example, the mass 693 a.m.u. is obtained from the ion with mass 648 a.m.u. by adding 5 x 9 a.m.u. resulting from d_9 groups, which indicates 5 TMS groups on this fragment. This spectrum is in agreement with the suggested fragmentation from Figure 5. A similar result as shown for the spectrum of epigallocatechin gallate, can be obtained for any other silylated compound.

Besides the similarity in the spectrum profile, the d_9-silylated compounds have a similar retention time as those silylated with non-isotopically labeled BSTFA, and the chromatogram also has a similar profile, as shown in Figure 7, that displays two time windows between 58 min and 80 min from a green tea water extract derivatized with BSTFA and for the same extract derivatized with d_9-BSTFA.

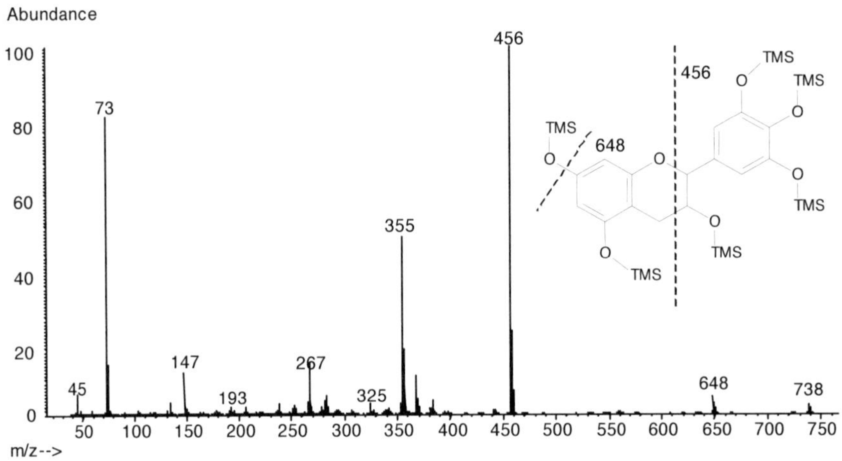

Figure 2. Spectrum of silylated epigallocatechin (EGC) (from standard compound), ret. time 65.18 min, MW = 738.31 (Note: The molecular weight does not consider the natural isotope distribution of elements and it is based only on the nuclidic masses of a single isotope).

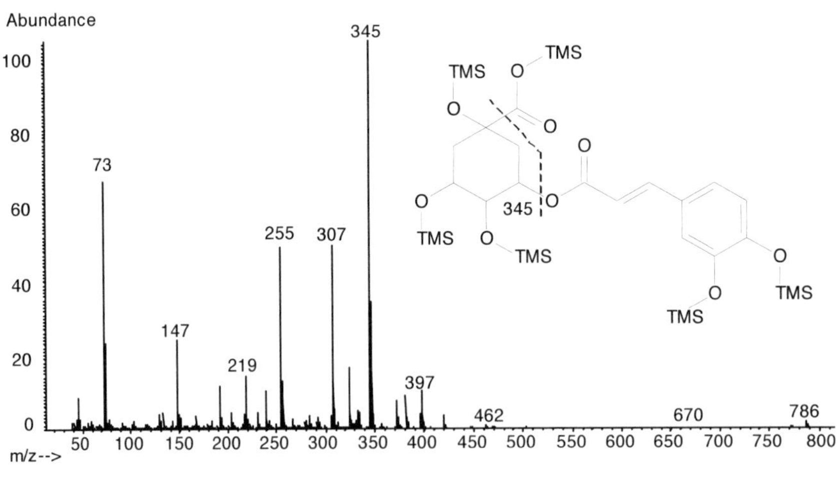

Figure 3. Spectrum of silylated chlorogenic acid (from standard compound), ret. time 68.11 min, MW = 786.33.

http://dx.doi.org/10.5772/57292

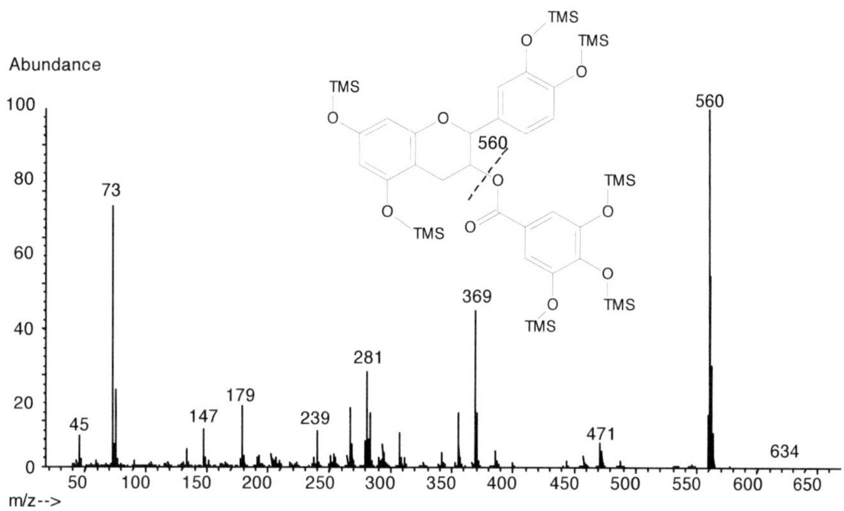

Figure 4. Spectrum of silylated epicatechin gallate (ECG) (from standard compound), ret. time 78.14 min, MW = 946.37.

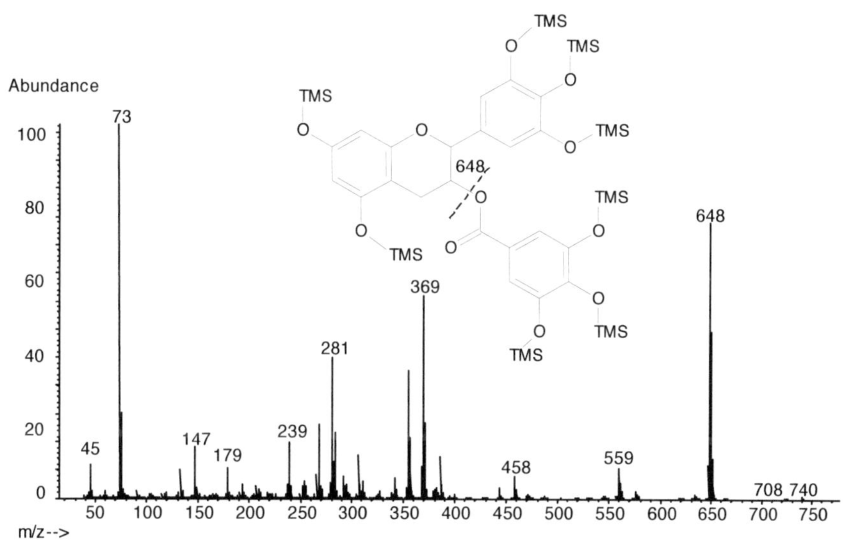

Figure 5. Spectrum of silylated epigallocatechin gallate (EGCG) (from standard compound), ret. time 78.83 min, MW = 1034.40. At higher instrument sensitivity, small peaks at 1034 a.m.u. and 1019 a.m.u. (loss of 15 a.m.u.) can be seen.

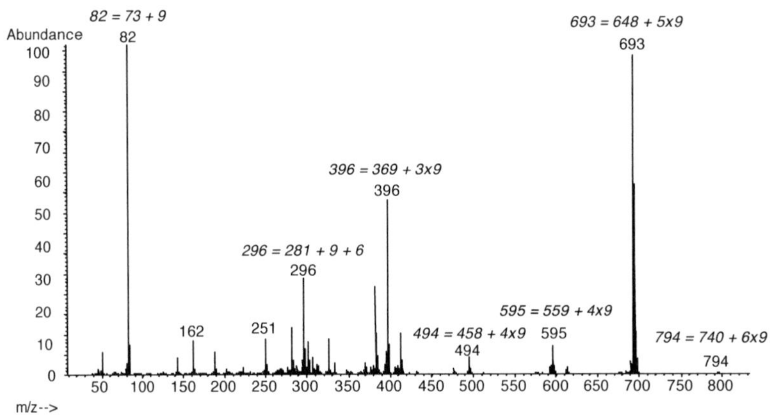

Figure 6. Spectrum of d_9-silylated epigallocatechin gallate. Masses of fragments are shown as related to those in the spectrum of non-deuterated compound displayed in Figure 5.

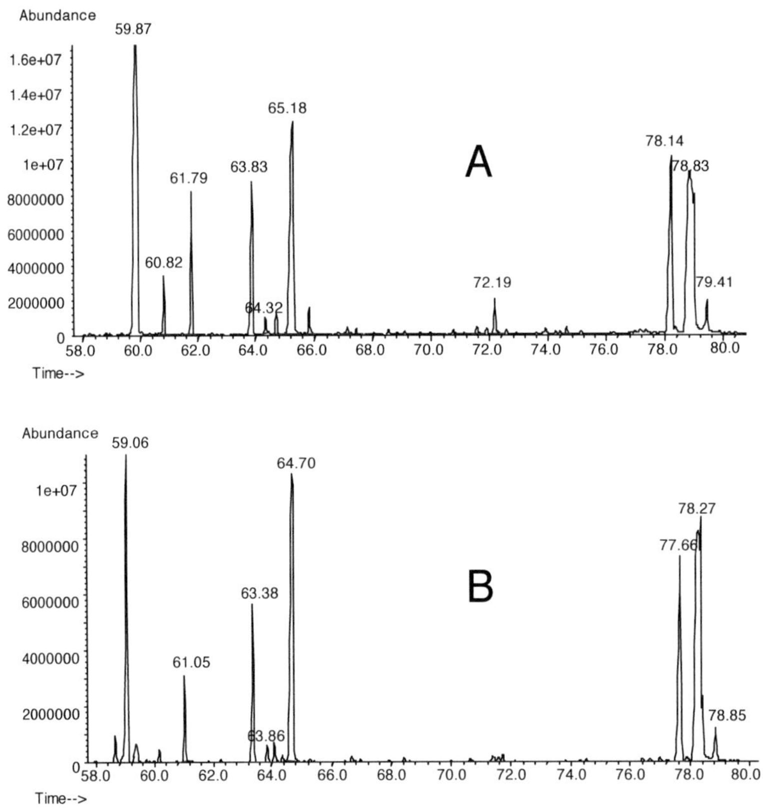

Figure 7. Two time windows between 58 min and 80 min from a green tea water extract derivatized with BSTFA (A) and for the same extract derivatized with d_9-BSTFA (B).

The quantitation of epigallocatechin and epigallocatechin gallate in the green tea was also evaluated in this study. For this purpose, the initial peak area was measured in the chromatogram of the silylated green tea sample. This was followed by the addition of 500 μg and 1000 μg of the two compounds (as solution in DMF) to 50 mg green tea sample with silylation. In order to avoid peak overloading, the silylated solution that was filtered through a 0.45 μm PTFE filter. The samples were analyzed by GC/MS and the peak areas were measured. The results are illustrated in Figure 8 that represents the peak area measurement normalized by the internal standard area (0.04 mg/mL *tert*-butylhydroquinone after 1/10 dilution) versus the compound addition.

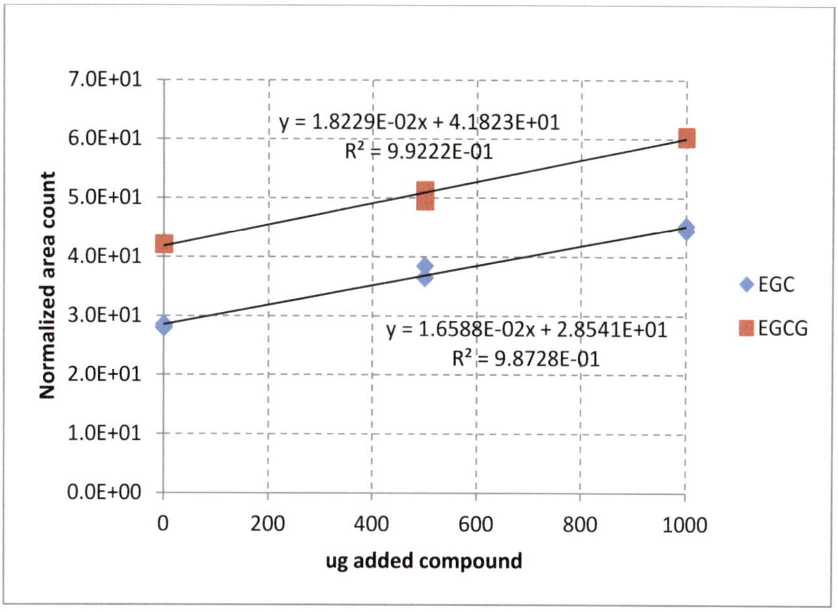

Figure 8. Peak area for epigallocatechin (EGC) and epigallocatechin gallate (EGCG) for a green tea sample, and for the same type of sample with 500 μg and 1000 μg of the two compounds added.

From the trendline equations, it can be calculated that the green tea contained about 2.307 mg EGCG/50 mg sample, and 1.720 mg EGC/50 mg sample. This is equivalent to 46.14 mg/g EGCG and 34.41 mg/g ECG. These levels are in the range reported in other studies for green tea [34]. Green tea from different sources may have different levels of antioxidants, and silylation followed by GC/MS analysis is an excellent tool for comparing these levels.

3.2. Example of rosemary analysis

Rosemary (dry leaf) (*Rosmarinus officinalis*) is another botanical with antioxidant properties [22,23]. For rosemary (dry leaf) evaluated in this study (commercially available from American Spice Trading Co.) the ORAC values (both lipophilic and hydrophilic) were 620 ± 20 μM TE/g, and FRAP was 800 ± 15 μM Fe^{2+}/g. The chemical composition of rosemary leaf can be studied using the GC/MS technique after direct silylation of the dry leaf similarly to green tea.

The chromatogram of silylated rosemary is shown in Figure 9, and the identification of the main peaks can be viewed in Table 3 where the retention times for individual compounds are listed.

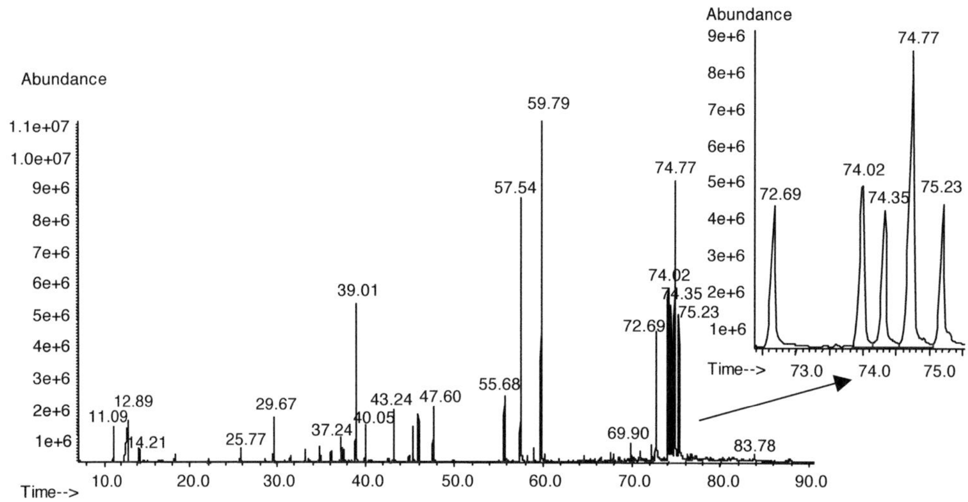

Figure 9. Chromatogram of silylated rosemary dry leaf. The identification of main peaks is given in Table 3. (I.S. elutes at 29.67 min).

Antioxidant compounds	**Ret. Time**	**Other main compounds**	**Ret. Time**
Catechol lactate	45.33	Camphor	11.09
Caffeic acid	47.60	Borneol	14.21
Rosmaricin	55.68	Malic acid	25.77
Carnosic acid	57.54	Pentose (ribose ?)	34.71
Carnosol	58.90	Fructose	37.24
Rosmanol	60.19	Quinic acid	39.01
Rosmarinic acid	72.69	Glucose	40.05, 43.24
Oleanolic acid	74.02	Myoinositol	45.91
Betulinic acid	74.35	Sucrose	59.79
Ursolic acid	74.77	Disaccharide	69.90
Betulonic acid	75.23	Disaccharide ?	83.78

Table 3. Compound identification for rosemary chromatogram shown in Figure 9

http://dx.doi.org/10.5772/57292

Some of the spectra of the silylated compounds are not available in common mass spectral libraries. The spectra of silylated rosmaricin, carnosic acid, carnosol, rosmanol, rosmarinic acid, oleanolic acid, betulinic acid, ursolic acid, and betulonic acid are shown in Figures 10 to 18.

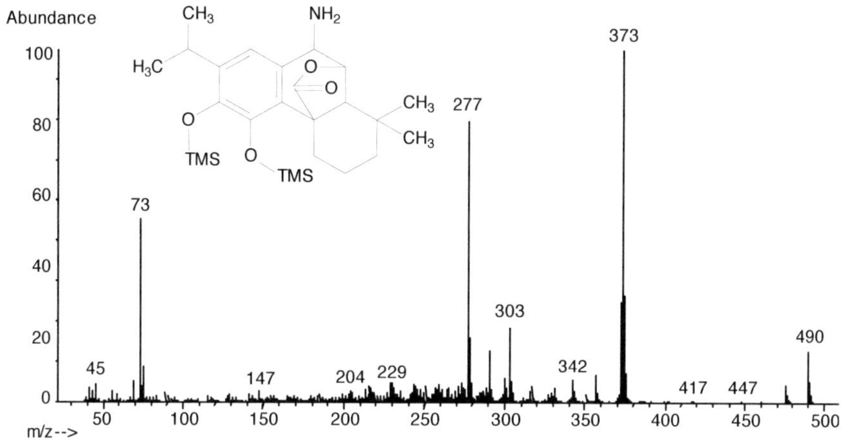

Figure 10. Spectrum of silylated rosmaircin, ret. time 55.68 min, MW = 489.27 (Note: the presence of two silyl groups on this molecule has been verified using silylation with d_9-BSTFA).

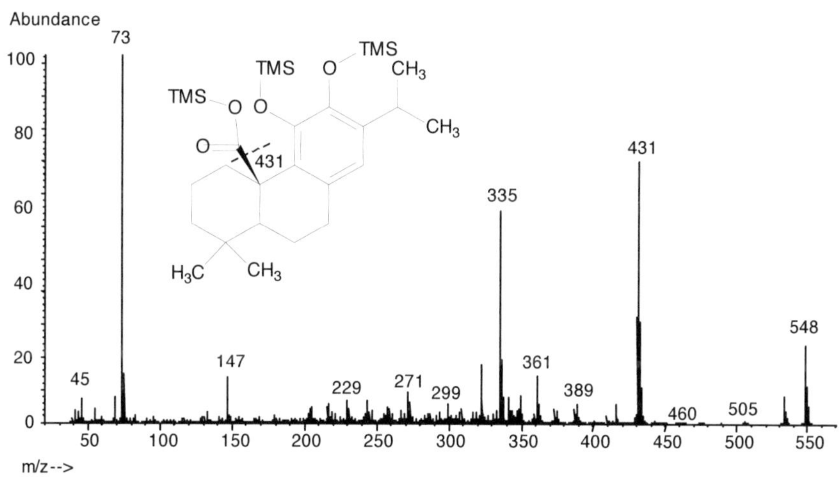

Figure 11. Spectrum of silylated carnosic acid, ret. time 57.54 min, MW = 548.32.

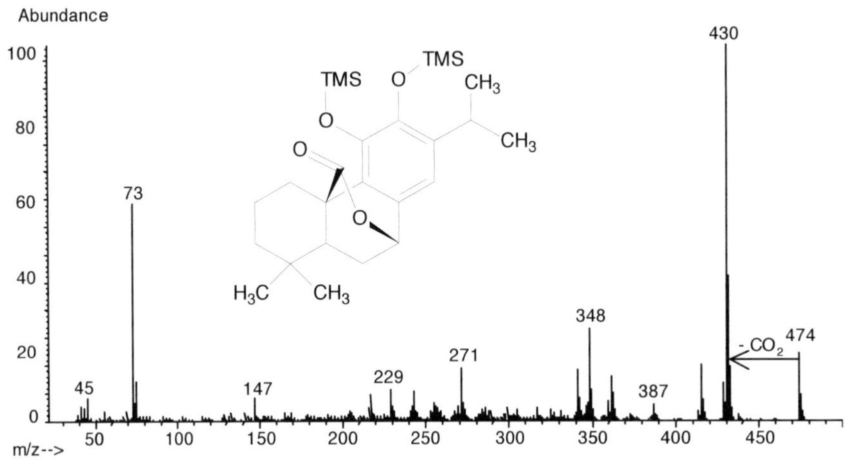

Figure 12. Spectrum of silylated carnosol, ret. time 58.90 min, MW = 474.26.

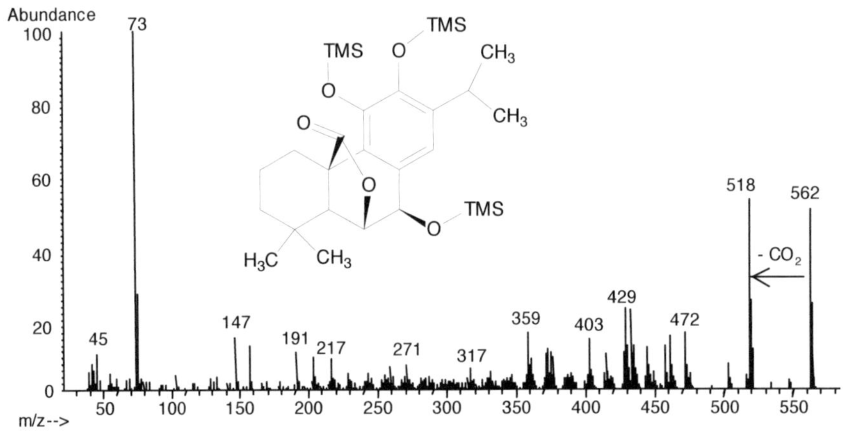

Figure 13. Spectrum of silylated rosmanol, ret. time 60.19 min, MW = 562.30.

http://dx.doi.org/10.5772/57292

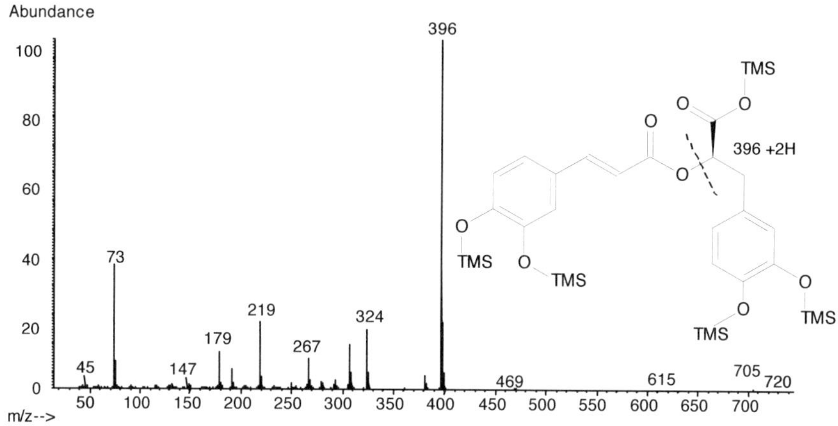

Figure 14. Spectrum of silylated rosmarinic acid, ret. time 72.69 min, MW = 720.28.

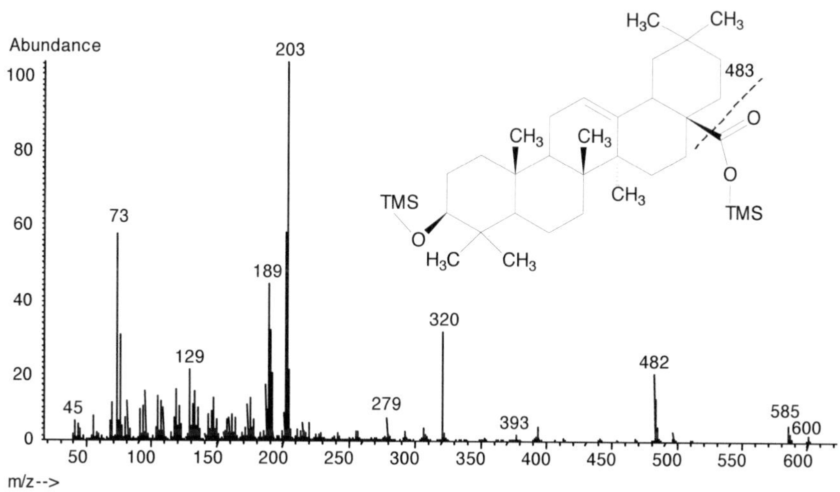

Figure 15. Spectrum of silylated oleanolic acid or 3β-hydroxyolean-12-en-28-oic acid (from standard), ret. time 74.02 min, MW = 600.44.

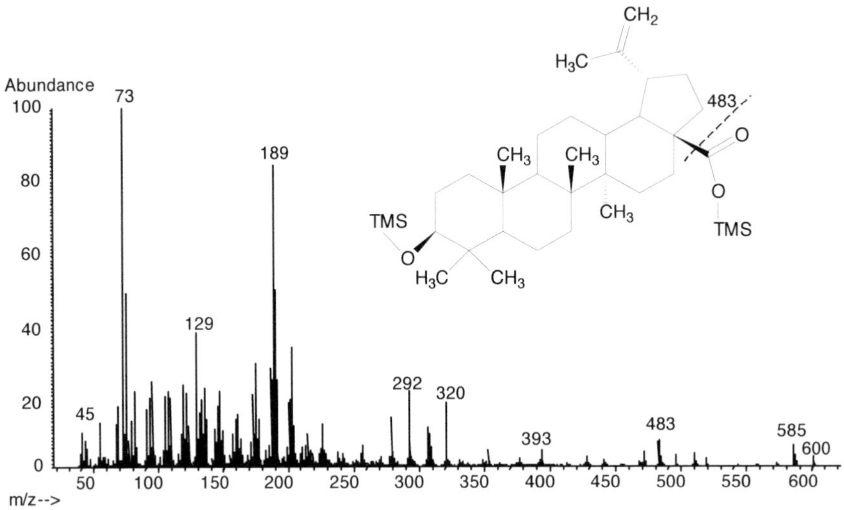

Figure 16. Spectrum of silylated betulinic acid, or (3β)-3-Hydroxy-lup-20(29)-en-28-oic acid (from standard), ret. time 74.35 min, MW = 600.44.

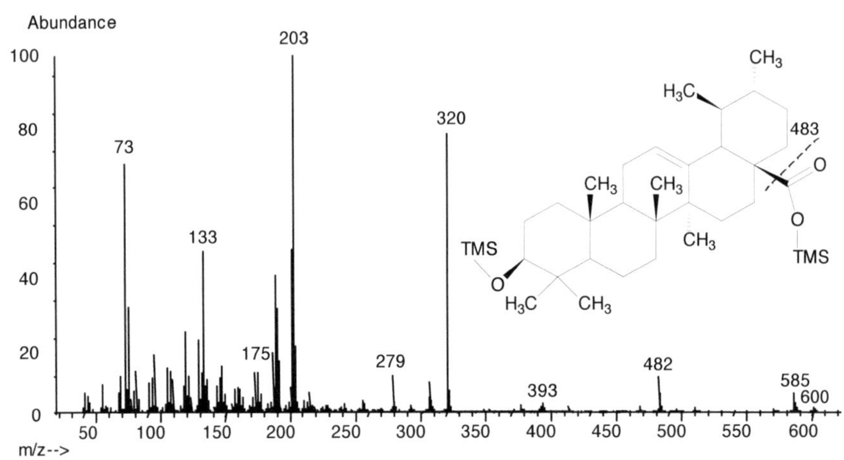

Figure 17. Spectrum of silylated ursolic acid or 3-β-hydroxy-urs-12-en-28-oic acid (from standard), ret. time 74.77 min, MW = 600.44.

http://dx.doi.org/10.5772/57292

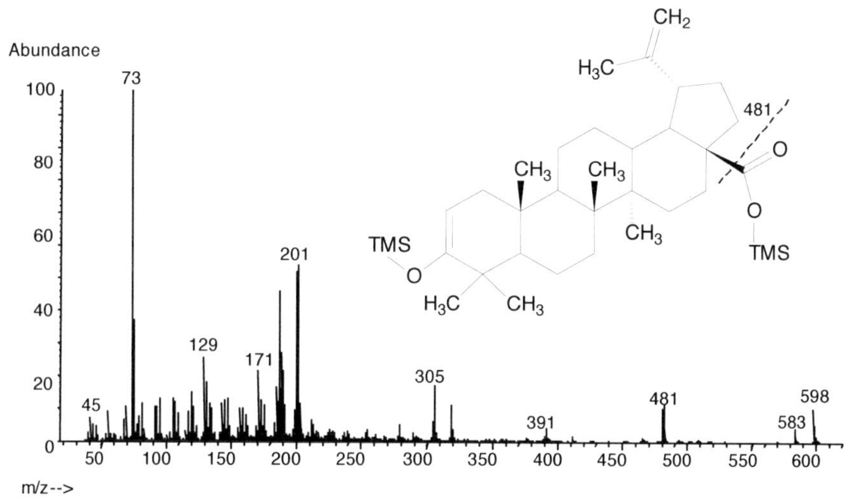

Figure 18. Spectrum (tentative) of silylated betulonic acid or 3-oxo-lup-20(29)-en-28-oic acid, ret. time 75.23 min, MW = 598.42.

The spectrum of silylated betulonic acid has a similar pattern to that of betulinic acid, except for several fragments being lower by two a.m.u. It was assumed that during silylation, the carbonyl group in position 3 is enolised and silylated.

The GC/MS analysis with direct derivatization of dry leaf of a botanical has various specific advantages compared to other analysis techniques. Besides its simplicity, the technique allows a detailed identification of the compounds seen in the chromatogram, allows a comparison of peak intensity between different types of botanicals, and quantitation when standards are available. An example of the application of this technique is the study of stability upon heating of rosemary regarding its antioxidant level. Starting from room temperature the heating was performed at three intervals up to 120 °C, for two hours. The variation in normalized area counts in the chromatograms of silylated leaf heated at different temperatures is shown in Figure 19. The results show that carnosic acid and rosmaricin have the tendency to decrease as the leaves are heated, while other antioxidant compounds are not affected by the heating in the indicated range.

3.3. Other applications of direct silylation and GC/MS analysis

A variety of other botanicals (leaves, rhizomes, or other plant parts) containing antioxidant molecules form silyl derivatives can be analyzed by GC/MS. Among the compounds that can be identified by sylilation and GC/MS are: vitexin, isoorientin, mangiferin, gossypin, delphinidin and other cyanidins, quercetin, tocoferol, coumaroyl quinic acid, ar-turmerone, curcumin, leucocyanidin gallate, etc. Some of the mass spectra of these molecules are easily identifiable, but in other cases, the identification is less obvious. In cases of glucosides (and C-glucosides), for example, the set of ions 147, 204, 217, 305 that are characteristic for the carbohydrate (glucose) moiety may lead to the conclusion that the chromatographic peak

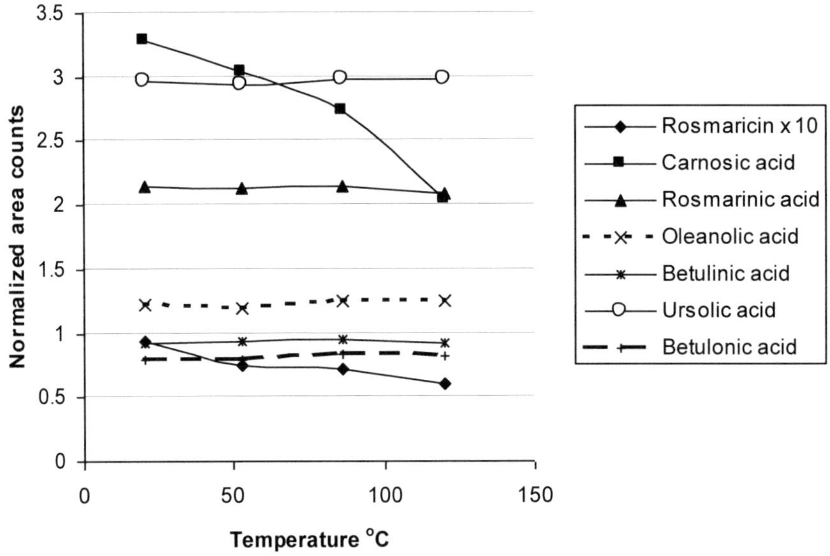

Figure 19. Variation with temperature in the normalized area counts of several antioxidant compounds (rosmaricin level is shown x 10 for being in scale) in rosemary heated between 20 °C and 120 °C, for two hours.

belongs to a carbohydrate, since carbohydrates are frequently present in plant extracts. As an example, the spectrum of silylated isoorientin (luteolin-6-C-glucoside) is given in Figure 20. In this spectrum, the presence of MW - 15 ion caused by the loss of a CH_3 from the silyl group, which is typical for silyl derivatives is a good indication of the parent molecule isoorientin which has MW =1024.41.

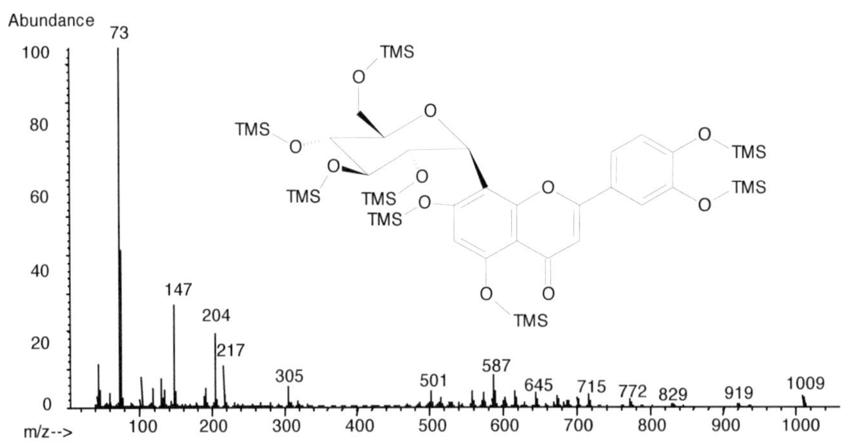

Figure 20. Spectrum of silylated isoorientin, (luteolin-6-C-glucoside) ret. time 82.68 min, MW = 1024.41. The peak with 1009 a.m.u. resulting from the loss of CH_3 is seen in the spectrum.

http://dx.doi.org/10.5772/57292

4. GC/MS analysis of triglycerides with antioxidant character

Some triglycerides present in botanicals, usually from the fruits or from seeds, are known to have antioxidant character. This character is caused by the presence of polyunsaturation in the long chain hydrocarbon moiety of the fatty acids (PUFAs) that are typically part of the triglyceride molecules. PUFAs (free or as triglyceride) have a scavenging potential toward reactive oxygen/nitrogen (ROS/RNS) species [36]. Several nomenclature systems are used for the fatty acids, a common one being omega-x (ω - x, or n - x). The value of x indicates the position of the double bond which is the closest to the terminal methyl of the hydrocarbon chain of the acid, with counting from the terminal methyl. For example, linoleic acid is a n - 6 or an omega-6 acid. Triglycerides formed from omega-3 acids, besides the antioxidant character, are considered essential fatty acids, since they cannot be synthesized by the human body and are related to additional health benefits. Common analysis of triglycerides is done either for the intact compound, or after hydrolysis and derivatization of the acid with methyl groups [37], or with silyl groups [29,38].

4.1. Triglyceride hydrolysis and fatty acids methylation

The formation of methyl esters from triglycerides is typically done in one operation that produces both the hydrolysis of the triglyceride and the methylation of the free acids formed in the hydrolysis. Methyl esters of the fatty acids (FAME) can be obtained using various reagents [30,39]. Common procedures use methanol and H_2SO_4, methanol and BF_3 [40-42] or methanol and HCl. One standard procedure [43] starts with the addition to 100 mg lipid in a 50 mL round bottom flask with condenser. To the flask is added 4 mL of a 0.5 M methanolic solution of NaOH. The solution is boiled until fat globules disappear. Then, 5 mL solution of BF_3 in methanol (125 g BF_3/L) is added and the boiling is continued for 2-3 min. Then about 5 mL of heptane is added and boiled for another minute. The mixture is allowed to cool and 15 mL saturated solution of NaCl is added. About 1 mL heptane is collected from the upper layer and is dried over anhydrous Na_2SO_4. The solution is diluted with heptane if necessary for the GC analysis. Detection for the GC can be either flame ionnization (FID) or mass spectrometry (MS). A number of variants of this methylation procedure are reported in the literature (e.g. [44]). For example, one variant starts with 200–500 mg lipid which is boiled with 5 mL 0.5 N NaOH or KOH in methanol for 3–5 min. To this mixture is added 15 mL of an esterification solution, and the mixture is refluxed for 3 min. The esterification solution is prepared by adding 2 g NH_4Cl to 60 mL methanol and 3 mL conc. H_2SO_4 which are than refluxed together for 15 min. The esterified acids are transferred into a separation funnel containing 25 mL petroleum ether and 50 mL water. The water is discarded and the organic phase is washed twice with 25 mL water. The resulting organic phase can be concentrated, dried with Na_2SO_4, and analyzed by GC. The reactions taking place are described as follows:

The analysis of FAME can be performed following various procedures. One such procedure uses a SP2560 100 m x 0.25 mm column with 0.2 μm film for separation. This is a highly polar biscyanopropyl column specifically designed to separate geometric position isomers of fatty acid methyl esters. The recommended GC conditions are given in Table 4.

Parameter	Description	Parameter	Description
Initial oven temp.	100°C	Injection volume	1.0 μL
Initial time	4.0 min	Carrier gas	Helium
Oven ramp rate	3°C/min	Flow mode	Constant flow
Oven final temp.	240°C	Flow rate	0.75 mL/min
Final time	15 min	Linear flow rate	18 cm/s
Total run time	65.6 min	Split ratio	200 : 1
Inlet temp.	225°C	GC outlet	FID
Inlet mode	Split	Detector temperature	300°C

Table 4. GC operating parameters for methyl ester analysis.

The procedure allows the separation of over 60 FAME. Other columns and shorter run times can be utilized if a less detailed separation is desired. For example, a SP2380 30 m x 0.25 mm column with 0.2 μm film can be used, with oven starting at 150 °C and gradient to 250 °C at 4 °C/min, helium carrier gas at 20 cm/s (at 150 °C), and FID detector at 260 °C. In these conditions, a typical GC/MS chromatogram obtained for linseed oil is shown in Figure 21.

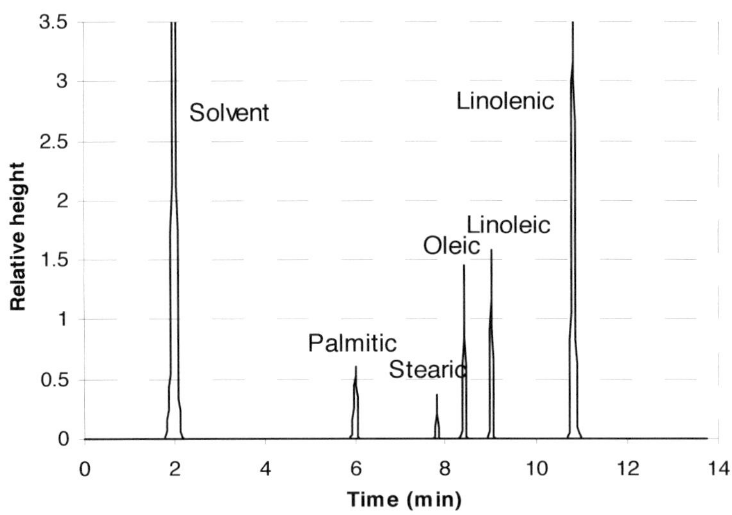

Figure 21. GC/MS chromatogram for methyl esters in linseed oil.

Other procedures to generate methyl esters are also reported in the literature [38].

4.2. Triglyceride hydrolysis and fatty acids silylation

Hydrolysis and formation of silyl derivatives of fatty acids is another procedure used for lipid analysis. The analysis starts with the hydrolysis of the triglycerides. For this purpose, 0.3 to 0.5 mg lipid (precisely weighed) was treated with 50 μL solution of 2M KOH in ethanol. The mixture was heated in a 2 mL capped vial for 30 min at 78 °C in a heating block, to generate potassium salts of the fatty acids. After that, the cap of the vial was removed, and the ethanol evaporated. Complete evaporation of ethanol, which takes 3-5 min, is necessary to avoid the formation of small proportions of ethyl esters when HCl is further added. To the vial, 25 μL solution of 6M HCl was added to neutralize the base and change the organic acid potassium salts into free acids. Then, 750 μL of *n*-nonane was added to extract the free acids. The nonane solution was treated in the vial with about 0.2 g of anhydrous Na_2SO_4 for drying. From the dry nonane solution, 500 μL were transferred into a separate 2 mL vial, treated with 25 μL of pyridine, 100 μL of dimethylformamide (DMF) that contains 400 μg/mL of tert-butylhydroquinone (TBHQ), and with 300 μL bis(trimethylsilyl)-trifluoroacetamide (BSTFA) with 1% trimethylchlorosilane (TMCS). The TBHQ is used as a chromatographic standard. The vials with the samples were heated at 78 °C for 30 min, followed by GC/MS analysis. The analysis of the samples was performed using a GC/MS instrument (Agilent 7890/5975 system, Wilmington, DE, USA), equipped with a Zebron ZB-50 column (Phenomenex, Torrence, CA 90501-1430, USA) that was 60 m long, 0.25 mm i.d., and 0.50 μm film thickness. The recommended parameters for the GC/MS analysis are given in Table 5

Parameter	Description	Parameter	Description
Initial oven temp.	50°C	Carrier gas	Hydrogen
Initial time	0.5 min	Flow mode	Constant flow
Oven first ramp rate	10°C/min	Flow rate	0.71 mL/min
Final oven temp. first ramp	200°C	Nominal initial pressure	12.05 psi
Final time first ramp	0 min	Split ratio	20 : 1
Oven second ramp rate	3°C/min	Split flow	14.20 mL/min
Final oven temp. second ramp	250°C	GC outlet	MSD
Final time second ramp	0 min	Outlet pressure	Vacuum
Oven third ramp rate	20°C/min	Transfer line heater	300°C
Final oven temp. third ramp	300°C	Ion source temp.	230°C
Final time third ramp	2 min	Quadrupole temp.	150°C
Total run time	36.66 min	MSD EM gain	2.0
Inlet temp.	300°C	MSD solvent delay	8.0 min
Inlet mode	Split	MSD acquisition mode	scan
Injection volume	0.5 μL	Mass range	33 to 550 a.m.u.

Table 5. GC/MS operating parameters for silylated acids analysis.

The peak identification was performed using both standards (when available) and mass spectra library searches (on NIST 08 library). The chromatography allows excellent separation of acids in the range C6 to C27, and differentiate isomers such as oleic and elaidic acid. Quantitation of fatty acids was obtained using calibration curves. A typical total ion chromatogram (TIC) for the fatty acids as TMS derivatives from a commercial vegetable cooking oil hydrolysate sample is shown in Figure 22.

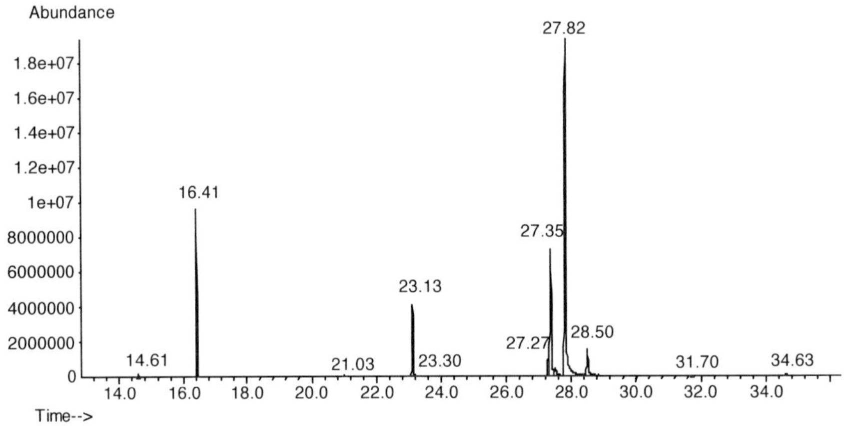

Figure 22. GC/MS chromatogram of TMS derivatives of fatty acids from a commercial cooking oil hydrolysate sample. Peak identification can be obtained using the data from Table 6.

No.	Compound	Ret. time	MW	Identifying ions	Formula	Symbol	Area %
1	Glycerin 3TMS (not shown)	10.74	308.64	205, 218	C12H32O3Si3		0.15
2	Unknown	14.61	?	192, 163	?		0.24
3	Internal standard (I.S.)	16.41					
4	Column bleed	21.03					
5	Palmitic acid TMS	23.13	328.613	313, 328	C19H40O2Si	C16:0	9.52
6	Palmitoleic acid TMS	23.30	326.597	311, 326	C19H38O2Si	C16:1 Z-9	0.07
7	Stearic acid TMS	27.27	356.667	341, 356	C21H44O2Si	C18:0	2.56
8	Oleic acid TMS	27.35	354.651	339, 354	C21H42O2Si	C18:1 Z-9	20.33
9	Elaidic acid TMS (trans-9-C18:1)	27.49	354.651	339, 354	C21H42O2Si	C18:1 E-9	2.09
10	Linoleic acid TMS	27.82	352.635	337, 352	C21H40O2Si	C18:2 Z,Z-9,12	59.72
11	Linolenic acid TMS	28.50	350.62	335, 350	C21H38O2Si	C18:3 Z,Z,Z-6,9,12	4.99
12	Arachidic acid TMS	31.70	384.721	369, 384	C23H48O2Si	C20:0	0.13
13	11-Eicosenoic acid TMS	31.80	382.705	367, 382	C23H46O2Si	C20:1 Z-11	0.06
14	Docosanoic acid TMS (behenic)	34.63	412.78	397, 412	C25H52O2Si	C22:0	0.14

Table 6. Peak identification and relative peak area for the chromatogram of TMS derivatives of fatty acids from a commercial vegetable cooking oil hydrolysate sample.

4.3. Analysis of intact triglycerides

For the analysis of triglycerides as whole molecules, a solution containing about 0.5 mg/mL lipid in n-nonane (b.p. 151 °C) was made from each sample. This solution was analyzed directly by GC, in conditions described in Table 7. The GC was equipped with a Rtx®-65TG column, 30 m x 0.25 mm, with 0.1 µm film thickness (Restek, Bellefonte, PA 16823, USA). Similar separation was obtained using a CP-Tap column, 25 m x 0.25 mm, 0.1 µm film (Varian, Walnut Creek, CA 94598, USA) in the same conditions as in Table 7. The GC can be used either with FID detection or MS detection. The conditions for the MS and FID detectors are shown in Table 8.

Parameter	Description	Parameter	Description
Initial oven temperature	130 °C	Inlet mode	Ramped
Initial time	1.0 min	Inlet initial temperature	130 °C
Oven temp. rate first ramp	30 °C/min	Initial time	0.1 min
Final temperature first ramp	300 °C	Inlet temperature rate	150 °C/min
Final time	0.0 min	Final temperature	300 °C
Oven temp. rate second ramp	4.0 °C/min	Injection volume	0.2 µL
Final temperature second ramp	365 °C	Carrier gas	H_2
Final time	7.0 min	Flow mode	Constant flow
Total run time	29.92 min	Flow rate	0.8 mL/min
Inlet	Cold on column		

Table 7. GC operating parameters.

MS Parameter	Description	FID Parameter	Description
MSD transfer line	300 °C	Detector temperature	300 °C
Ion source temperature	230 °C	H_2 flow	30 mL/min
MSD EM gain	2.0	Air flow	400 mL/min
MSD solvent delay	3.0 min	Make up flow N_2	25 mL
MS operating mode	Scan EI+		
Mass range a.m.u.	50 – 800 a.m.u.		

Table 8. ID and MS operating parameters.

Using the conditions previously described, the chromatogram of a commercial vegetable cooking oil with FID detection is shown in Figure 23, and with MS detection is shown in Figure 24.

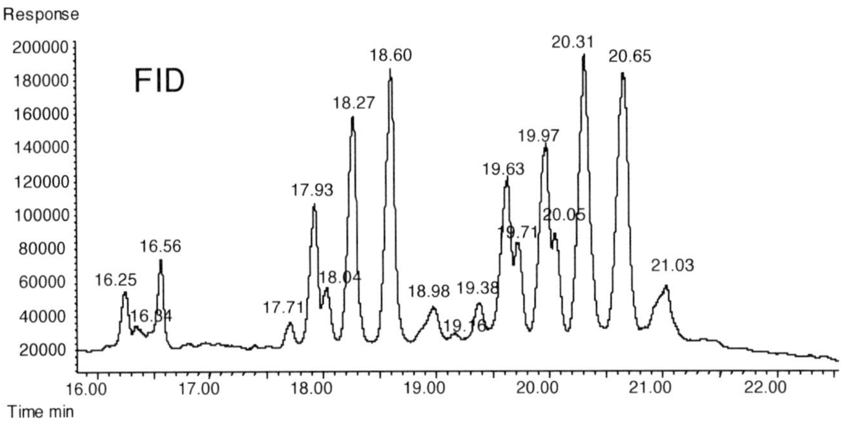

Figure 23. Chromatogram of a commercial vegetable cooking oil generated using FID detection.

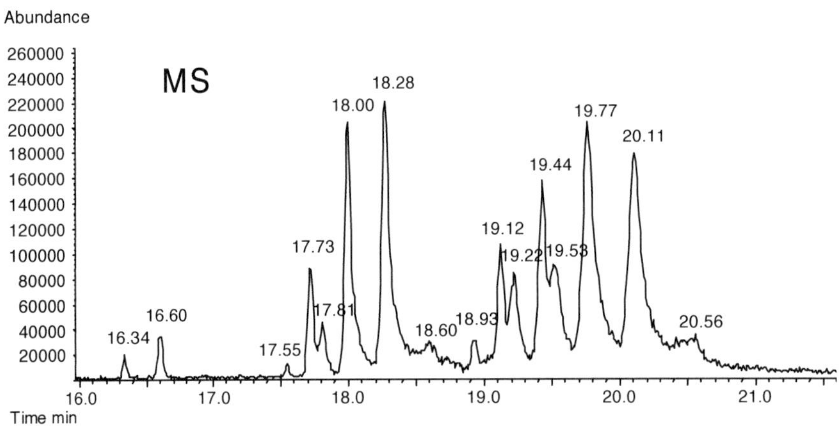

Figure 24. Chromatogram of a commercial vegetable cooking oil generated using MS detection (total ion chromatogram or TIC). Peak identification following retention times as given in Table 9.

http://dx.doi.org/10.5772/57292

	Compound	Formula	Ret. time in MS	MW	Identifying ions	Area % from MS	Triglyc. % from FID
1	Dipalmitin olein	C53H100O6	16.34	833.380	551, 577, 339	0.51	0.27
2	Dipalmitin linolein	C53H98O6	16.60	831.364	551, 575, 335	1.42	0.85
3	Palmitin stearin olein	C55H104O6	17.55	861.434	579, 605, 341	0.34	0.19
4	Palmitin diolein	C55H102O6	17.73	859.418	577, 603, 339	4.14	2.65
5	Palmitin stearin linolein	C55H102O6	17.81	859.418	579, 603, 341	2.14	1.37
6	Palmitin olein linolein	C55H100O6	18.00	857.402	577, 601, 339	12.51	9.19
7	Palmitin dilinolein	C55H98O6	18.28	855.386	575, 599, 337	15.93	13.44
8	Palmitin linolein linolenin	C55H96O6	18.61	853.370	573, 575, 335,	1.40	1.36
9	Linolein distearin	C57H106O6	18.93	887.472	605, 341, 264	1.03	0.70
10	Triolein	C57H104O6	19.13	885.456	603, 339, 264	5.34	4.56
11	Distearin olein	C57H108O6	19.23	889.488	605, 341, 262	5.31	3.17
12	Diolein linolein	C57H102O6	19.44	883.440	603, 339, 262	8.36	8.63
13	Stearin olein linolein	C57H104O6	19.53	885.456	603, 341, 262	8.68	6.87
14	Dilinolein olein	C57H100O6	19.77	881.424	601, 339, 262	16.37	20.21
15	Trilinolein	C57H98O6	20.11	879.408	599, 337,262	12.90	19.39
16	Dilinolein linolenin	C57H96O6	20.53	877.392	597, 599, 337	3.62	6.59

Table 9. Peak identification, relative peak area for the MS chromatogram and % triglyceride from FID measurement for a commercial vegetable cooking oil.

The peak identifications can be done based on the mass spectra of each compound. For example, the mass spectrum of palmito-linoleo-olein is given in Figure 25.

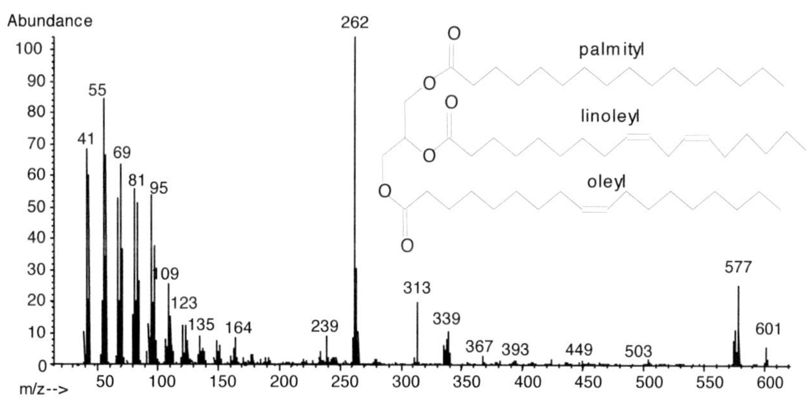

Figure 25. Mass spectrum of palmito-linoleo-olein (the correct position of substituents is unknown).

The structures of several diagnostic ions in the spectrum of a triglyceride species that contains palmityl, linoleyl, and oleyl fatty acids in the molecule is given in Figure 26.

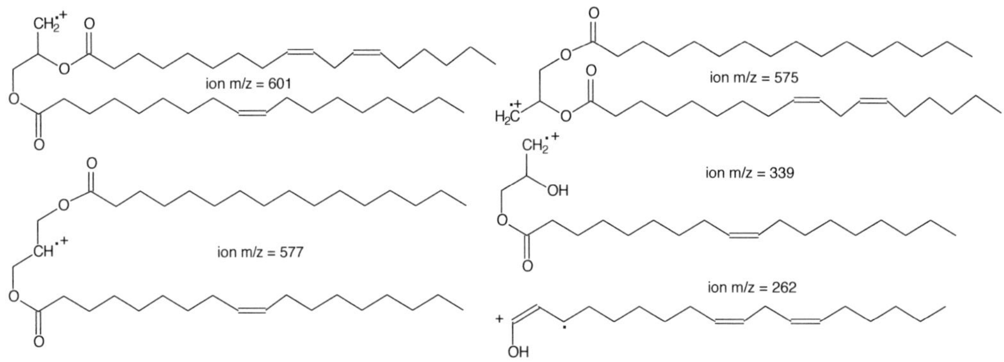

Figure 26. The structures of several diagnostic ions in the spectrum of palmito-linoleo-olein.

Heavier triglycerides are less amenable for direct GC analysis. For example, direct measurement of triglycerides esterified with more than two linolenic acids is not possible in the chromatographic conditions previously described. As an example, the TIC trace for a sample of linseed oil generated in the same conditions as the chromatogram from Figure 25 is given in Figure 27 [29].

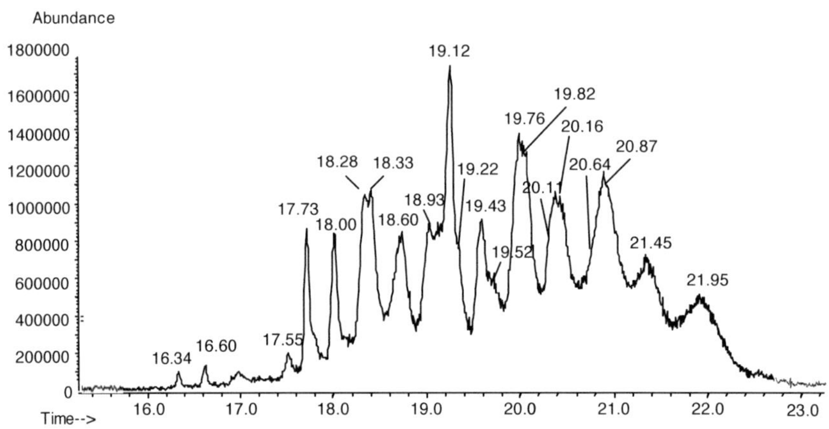

Figure 27. Chromatogram of linseed oil generated using MS detection. Peak identification following retention times as given in Table 9 and in Table 10.

Many peaks from this chromatogram are identical to those described in Table 9. However, a few additional triglycerides were identified (some tentatively) in linseed oil and they are given in Table 10.

	Compound	Formula	Ret. time in MS	MW	Identifying ions
1	Palmito oleino linolenin	C55H98O6	18.33	855.386	577, 599, 573
2	Dioleino linolenin?	C57H100O6	19.82	881.424	604, 599, 339
3	Stearo linoleo linolenin	C57H100O6	20.28	881.424	603, 601, 597
4	Oleino linoleo linolenin	C57H98O6	21.45	879.408	597, 599, 601
5	Oleino dilinolenin	C57H96O6	21.95	877.392	595, 599, 335

Note: Ions with m/z value in bold result from the loss of a fatty acid residue from the triglyceride molecule.

Table 10. Peak Identification, for Extra Peaks in the MS Chromatogram for Linseed Oil.

5. Conclusions

GC/MS is a very useful technique for the analysis of antioxidants in botanicals, although many antioxidant molecules are large and/or contain numerous polar groups. GC methods have limitations regarding their capability to be used for the analysis of heavier and less volatile molecules. However, the use of derivatization of the analytes, and special selection of the GC settings allow the extension of the applicability for this technique. The unique capability to identify molecular species based on EI+ mass spectra makes GC/MS an invaluable tool in the analysis of antioxidants in botanicals.

Author details

Serban Moldoveanu*

Address all correspondence to: moldovs@rjrt.com

R.J. Reynolds Tobacco Co., Winston-Salem, USA

References

[1] D. Huang, B. Ou, R. L. Prior, The chemistry behind antioxidant capacity assays, *J. Agric. Food Chem.*, 53 (2005) 1841-1856.

[2] H. Wang, G. Cao, R. L. Prior, Total antioxidant capacity of fruits, *J. Agric. Food Chem.*, 44 (1996) 701-705.

[3] S. M. Henning, C. Fajardo-Lira, H. W. Lee, A. A. Youssefian, V. L. W. Go, D. Heber, Catechin content of 18 teas and a green tea extract supplement correlates with the antioxidant capacity, *Nutrition and Cancer*, 45 (2003) 226–235.

[4] European Food Safety Authority, Use of rosemary extracts as a food additive, *The EFSA Journal*, 721 (2008) 1-29.

[5] D. Del Rio, A. Rodriguez-Mateos, J. P.E. Spencer, M. Tognolini, G. Borges, A. Crozier, Dietary (poly)phenolics in human health: structures, bioavailability, and evidence of protective effects against chronic diseases, Antioxidants & Redox Signaling, 18 (2013) 1818-1892.

[6] G. Cao, H. M. Alessio, R. G. Cutler, Oxygen-radical absorbance capacity assay for antioxidants, *Free Radical Biology & Medicine*, 14 (1993) 303-311.

[7] G. Cao, A. H.Wu, H. Wang, R. L. Prior, Automated assay of oxygen radical absorbance capacity with the Cobas Fara II, *Clin. Chem.* 41 (1995) 1738–1744.

[8] B. Ou, M. Hampsch-Woodill, R. L. Prior, Development and validation of an improved oxygen radical absorbance capacity assay using fluorescein as the fluorescent probe., *J. Agric. Food Chem.*, 49 (2001) 4619–4626.

[9] I. F. F. Benzie, J. J. Strain, The ferric reduction ability of plasma (FRAP) as a measure of "antioxidant power": the FRAP assay, *Anal. Biochem.*, 239 (1996) 70-76.

[10] O. Folin, V. Ciocalteu, On tyrosine and tryptophane determinations in proteins, *J. Biol. Chem.*, 73 (1927) 627–650.

[11] A. Khafif, S. P. Schantz, T.-C. Chou, D. Edelstein, P. G.Sacks, Quantitation of chemopreventive synergism between (–)-epigallocatechin-3-gallate and curcumin in normal, premalignant and malignant human oral epithelial cells, *Carcinogenesis*, 19 (1998) 419–424.

[12] D. Del Rio, A.J. Stewart, W. Mullen, J. Burns, M. E. Lean, F. Brighenti, A. Crozier, HPLC-MS analysis of phenolic compounds and purine alkaloids in green and black tea. *J. Agric. Food Chem.*, 52 (2004) 2807–2815.

[13] M. N. Irakli, V. F. Samanidou, I. N. Papadoyannis, Development and validation of an HPLC method for the simultaneous determination of tocopherols, tocotrienols and carotenoids in cereals after solid-phase extraction, *J. Sep. Sci.* 34 (2011)1375–1382.

[14] S. Y. Li, Y. Yu, S. P. Li, Identification of antioxidants in essential oil of radix Angelicae sinensis using HPLC coupled with DAD-MS and ABTS-based assay, *J. Agric. Food Chem.*, 55 (2007) 3358–3362.

[15] D. Han, K. H. Row, Determination of luteolin and apigenin in celery using ultrasonic-assisted extraction based on aqueous solution of ionic liquid coupled with HPLC quantification, *J. Sci. Food Agric.*, 91 (2011) 2888–2892.

[16] M. Parvu, A. Toiu, L. Vlase, P. E. Alina, Determination of some polyphenolic compounds from Allium species by HPLC-UV-MS, *Nat. Prod. Res.* 24 (2010) 1318–2134.

[17] T. M. Rababah, K. I. Ereifej, R. B. Esoh, M. H. Al-u'datt, M. A. Alrababah, W. Yang, Antioxidant activities, total phenolics and HPLC analyses of the phenolic compounds of extracts from common Mediterranean plants, *Nat. Prod. Res.* 25 (2011) 596–605.

[18] C. Proestos, N. Chorianopoulos, G. J. Nychas, M. Komaitis, RP-HPLC analysis of the phenolic compounds of plant extracts. investigation of their antioxidant capacity and antimicrobial activity. *J. Agric. Food Chem.*, 53 (2005) 1190–1195.

[19] M. M. Naidu, B. N. Shyamala, J. R. Manjunatha, G. Sulochanamma, P. Srinivas, Simple HPLC method for resolution of curcuminoids with antioxidant potential, *J. Food Sci.*, 74 (2009) C312–C318.

[20] M. Reto, M. E. Figueira, H. M. Filipe, C. M. M. Almeida, Chemical composition of green tea (Camellia sinensis) infusions commercialized in Portugal, *Plant Foods Hum. Nutr.*, 62 (2007) 139–144.

[21] S. M. Henning, C. Fajardo-Lira, H. W. Lee, A. A. Youssefian, V. L. W. Go, D. Heber, Catechin content of 18 teas and a green tea extract supplement correlates with the antioxidant capacity, *Nutrition and Cancer*, 45 (2003) 226–235.

[22] F.J. Señoráns, E. Ibañez, S. Cavero, J. Tabera, G. Reglero, Liquid chromatographic–mass spectrometric analysis of supercritical-fluid extracts of rosemary plants, *J. Chromatog. A*, 870 (2000) 491–499.

[23] L. Almela, B. Sánchez-Muñoz, J. A. Fernández-López, M. J. Roca, V. Rabe, Liquid chromatograpic–mass spectrometric analysis of phenolics and free radical scavenging activity of rosemary extract from different raw material, J. Chromatog. A, 1120 (2006) 221-229.

[24] M. Herrero, M. Plaza, A. Cifuentes, E. Ibáñez, Green processes for extraction of bioactives from rosemary. Chemical and functional characterization via UPLC-MS/MS and in-vitro assays, *J. Chromatog. A*, 1217 (2009) 2512-2520.

[25] C. D. Stalikas, Extraction, separation, and detection methods for phenolic acids and flavonoids, (Review), J. *Sep. Sci.*, 30 (2007) 3268-3295.

[26] R. J. Robbins, phenolic acids in foods: an overview of analytical methodology, *J. Agric. Food Chem.*, 51 (2003) 2866-2887.

[27] I. Arslan, A. Celik, Chemical composition and antistaphylococcal activity of an endemic Salvia chrysophylla stapf. naturally distributed in Denizli province (Turkey) and its vicinity, *Pak. J. Bot.*, 40 (2008) 1799-1804.

[28] Z. Penton, Determination of neutral lipids by high temperature GC, Part I. Triglycerides and cholesteryl esters in foods, *Varian Application Note*, No. 22b.

[29] S. C. Moldoveanu, Y. Chang, Dual analysis of triglycerides from certain common lipids and seed extracts, J. *Agric. Food Chem.*, 59 (2011) 2137-2147.

[30] S. C. Moldoveanu, V. David, *Sample Preparation in Chromatography*, Elsevier, Amsterdam, 2002.

[31] F. Shahidi, C. -T. Ho (eds.), *Antioxidant Measurement and Applications*, ACS Symposium Ser. 956, ACS, Washington, 2007.

[32] D. J. Charles, *Antioxidant Properties of Spices, Herbs and Other Sources*, Springer, New York, 2013.

[33] S. M. Henning, C. Fajardo-Lira, H. W. Lee, A. A. Youssefian, V. L. W. Go, D. Heber, Catechin content of 18 teas and a green tea extract supplement correlates with the antioxidant capacity, *Nutrition and Cancer*, 45 (2003) 226–235.

[34] M. Reto, M. E. Figueira, H. Mota Filipe, C. M. M. Almeida, Chemical composition of green tea (*Camellia sinensis*) infusions commercialized in Portugal, *Plant Foods Hum. Nutr.* 62 (2007) 139–144.

[35] H. N. Graham, Green tea composition, consumption, and polyphenol chemistry, *Preventive Med.* 21 (1992) 334-350.

[36] D. Richard, K. Kefi, U. Barbe, P. Bausero, F. Visioli, Polyunsaturated fatty acids as antioxidants, *Pharmacol. Res.*, 57 (2008) 451-455.

[37] W. W. Christie, http://lipidlibrary.aocs.org/topics/ester_93/refs.htm

[38] S.C. Moldoveanu, Profiling of lipids from fruit and seed extracts, pp. 73-123, in Su Chen (ed.) *Lipidomics: Sea Food, Marine Based Dietary Supplement, Fruit and Seed*, Transworld Res. Network, Kerala, India, 2012.

[39] D. R. Knapp, *Handbook of Analytical Derivatization Reactions*, J. Wiley, New York, 1979.

[40] D. E. Albertyn, C. D. Bannon, J. D. Craske, N. T. Hai, N. L. Harper, K. L. O'Rourke, C. Szonyi, Analysis of fatty acids methyl esters with high accuracy and reliability I Optimization of flame-ionization detectors with respect of linearity, *J. Chromatogr.*, 247 (1982) 47-61.

[41] C. D. Bannon, J. D. Craske, N. T. Hai, N. L. Harper, K. L. O'Rourke, Analysis of fatty acids methyl esters with high accuracy and reliability II Methylation of fats and oils with boron trifluoride methanol, *J. Chromatogr.*, 247 (1982) 63-69.

[42] C. D. Bannon, G. J. Breen, J. D. Craske, N. T. Hai, N. L. Harper, K. L. O'Rourke, Analysis of fatty acids methyl esters with high accuracy and reliability III Literature review of and investigation into the development of rapid procedures for the methoxide-catalysed methanolysis of fats and oils, *J. Chromatogr.*, 247 (1982) 71-89.

[43] AOAC official method 969.22. Fatty acids in oils and fats. Preparation of methyl esters boron trifluoride method first action 1969

http://dx.doi.org/10.5772/57292

[44] AOAC Official method 996.06 Fat (Total, Saturated, and Unsaturated) in Foods.

Chapter 6

Influence of Culture Conditions on the Fatty Acids Composition of Green and Purple Photosynthetic Sulphur Bacteria

María Teresa Núñez-Cardona

Additional information is available at the end of the chapter

http://dx.doi.org/10.5772/57389

1. Introduction

As was published in [1], based on 16S rRNA sequences, prokaryotes are divided into two primary groups: Archaebacteria (Archaea) and Eubacteria (Bacteria). With the exception of Cyanobacteria, all the main lineages of photosynthetic organisms belong to the Eubacterial Phyla and the unrelated halobacterial species.

Within Eubacteria, photosynthetic organisms are found according to the following divisions: (i) gram-positive bacteria belonging to the family Heliobacteriaceae, (ii) green non-sulphur bacteria (also known as filamentous green bacteria) such as *Chloroflexus aurantiacus,* (iii) green sulphur bacteria such as *Chlorobium tepidum,* (iv) cyanobacteria, (v) and several genera of proteobacteria (or purple bacteria) [1].

Anoxygenic green and purple photosynthetic sulphur bacteria (GPSB/PPSB) are anaerobes. They do not release O_2 by photosynthesis, and as electron donors for this process, they use sulphur and its derivatives. These microorganisms have bacteriochlorophyll (Bchl) and accessory carotene pigments that include spirilloxanthin, okenone and chlorobactene pigment series.

Photosynthetic pigments in green photosynthetic sulphur bacteria (GPSB) include Bchl c, d and e, as well as chlorobactene and isorenieratene. PPSB may contain Bchl *a* or *b*. In general, these photosynthetic microorganisms present one kind of Bchl, except in GPSB - which could have small quantities of Bchl *a* - they belong to the Chlorobiaceae family. PPSB comprise the families Chromatiaceae (e.g., *Chromatium or Halochromatium)* and Ectothiorhodospiraceae.

Morphological patterns are limited in bacteria, and phenotypical together with molecular properties have been used for the identification of these microorganisms. Chemotaxonomy takes into account fingerprint methods using proteins, nucleic acids, sugars isoprenoids and fatty acids, among other molecules. For example, an average of fatty acids can be easily obtained by gas chromatography (or by gas-liquid chromatography). This average is called a 'fatty acid profile' and a part of this profile is a pattern [2]. The pattern is generated by cleaning and adapting the fatty acid profile. Both are highly reproducible when growth conditions, the physiological age of the cells and analysis are well standardized. Subsequently, the fatty acid patterns can be used for the identification of bacteria, library generation, taxonomy, epidemiological and ecological studies [2], and obviously it is very useful for detecting bacteria in samples of different origins.

Photosynthetic prokaryotes have a great variety of fatty acids, and they are useful for distinguishing these microorganisms; actually, they are recommended for the description and identification of new species ([3, 4]). For example, *Chloroflexus aurantiacus*, a green photosynthetic non-sulphur bacterium (GPNSB), contains straight-chain, saturated and monounsaturated fatty acids as its main constituents, but its pattern is distinctly different from the pattern of GSPB [5].

Fatty acids from whole cells of *Chlorobium* are within the range of C12-C18, and the main ones are: n-tetradecanoic (C14:0), hexadecenoic (C16:1) and n-hexadecanoic (C16:0) [5-7].

Currently, studies have reported 12 fatty acids in complete cells from GPSB, and the fatty acid C14:0 accounts for 14% [6] to 27.10% [5]; the fatty acid C16:1 accounts for at least 37.3% in these bacteria [5].

Hexadecanoic is the third main fatty acid in GPSB [6] in quantities from 10% in *Cb limicola v. thiosulfatophilum* to 29% in *Cb. phaeovibroides*; finally, the fatty acid C18:1 could be the other main one in GPSB, where it has been detected in quantities between 2.9% and 10.33% in the genus *Chlorobium* [5-7].

Minor fatty acids have been reported in these microorganisms and they include: C12:0, 12-Me-C14:0, 5-Me-C16:0, 14-Me-C16:0, C15:0, 15-Me-C16:1, C17:0, C18:0 and a cyclic fatty acid [5-7]. GPNSB (like *Chloroflexus*) are characterized by the presence of C17 and C18-C20 as the main fatty acids, and these are not detected in GPSB [5].

In oxygenic photosynthetic cyanobacteria - for example, in *Nostoc* and *Anabaena* - it has been reported [8] the presence of 17 major fatty acids (≥0.9%) that included saturated even-carbon, straight-chain C16:0 and C18:0; the first one was higher for *Anabaena* (27.7-34.72%) and lower for *Nostoc* isolates (18.5 to 26.1%). 26 minor components (at least 0.03%) and 10 trace-level components (average ≤ 0.03%) were also recorded.

Fatty acids of the Chromatiaceae family have been more studied than those from GPSB, and analysed strains of this family include members of the genera *Amoebobacter, Halochromatium* (e.g. *Chromatium salexigens*), *Thiocapsa, Allochromatium* (*Allochromatium vinosum* = *Chromatium vinosum*) and *Thiocystis.* As gram-negative bacteria, they have as main fatty acids: C18:1ω7>C16:1>C16:0. The proportions of these compounds are variable among species of these genera; however, C18:1 may be present with more than 50% (e.g., *Chromatium salexigens* DSM4395). The C16:1 contributes at least with 16 % (e. g., *Chromatium tepidum,* DSM 3771) and 36% as maximum (e. g., *Thiocystis gelatinosa*) [cited in 7], and the C16:0 fatty acid has been reported in quantities of 10.4% in *Chromatium* sp. (strain T7s9) to 25.8% in *Chromatium tepidum* (DSM 3771); it is important to note that in *Chr. tepidum,* the main fatty acids C18:1, C16:0 and C16:1 contribute with 46.38%, 25.86% and 16.17%, respectively, so the pattern for this strain is C18:1>C16:0>C16:1. Other fatty acids detected in minor quantities in members of the Chromatiaceae family include: C12:0, C13:0, C14:0, aC17:0, C18:1w9. C17:0, C18:0 and C20:1 [7-9].

The other family of PPSB, the Ectothiorhodospiraceae, has C18:1 as the main fatty acid, which contributes with 74.7% of the total extracted (96.7%) in *Ectothiorhodospira shaposhnikovii* [cited in 7]. The fatty acid C16:0 is the next in abundance, and C16:1 or C18:0 might be the third. The relative abundance of these major fatty acids as well as the less abundant are very different among the strains of this family (e.g., *Ec. halophila, Ec. halochloris, Ec. abdelmaleki, Ec. shaposhnikovii, Ec. vacuolata, Ec. salini, Ec. magna, Halorhodospira halophila* and *Thiorhodospira sibirica*) [cited in 7]. These minor fatty acids are: C12:0 [10] in *Ec. shaposhnikovii*; C19:0d8,9 (a phospholipid detected in *Halorhodospira halophila*) and 19:0Cy (detected in different strains of *Ectothiorhodospira halophila, Ec. Mobilis, Ec. salini* and *Ec. Abdelmaleki*). Also, C21:0 has been detected in the phospholipids of *Ec. shaposhnikovii* and *Halorhodospira halophila* DSM 244, while the eicosanoic (C20:0) fatty acid was found in *Halorhodospira halophila* BN9626 [cited in 7].

This chapter describes the influence of $MgSO_4$ and NaCl on three cultures of GPSB (*Chlorobium* genus) as well as the effect of culture age (15-30 days) in two strains of *Chromatium* (*Halochromatium*). Some considerations about the uses of fatty acids other than chemotaxonomy as well as new applications of fatty acid analyses like lipidomics by mass spectrometry are also included.

2. Fatty acids analysis

The analysis of fatty acids as methyl esters has been used to identify bacteria species. Their composition and the ratios of these compounds are useful in identifying microorganisms. They are conserved if bacteria are grown in similar conditions [11]. Also, it has been pointed out that information such as phylogenetic data and similarities derived from the fatty acid analysis are in broad agreement with 16S rDNA results, and appear to accurately distinguish between most species [12].

The number of bacterial species evaluated is a limitation in comparing the information available about the fatty acid composition from bacteria [12]. GPSB and PPSB are also not an exception compared to the fatty acid studies on chemoheterotrophic bacteria. There are very few investigations about the influence on fatty acid compositions of the culture medium for bacterial growth (e.g., salinity), the age of the cells or the physical factors (e.g., temperature) used to culture the microorganisms.

Fatty acid analysis may also depend on the extraction technique; in some studies, different procedures have been applied to investigate the fatty acid compositions of *Pseudomonas diminuta* and *P. maltophilia,* and the results have revealed that the qualitative and quantitative features of these compounds are different depending on the techniques used. Nevertheless, there are no changes in their profiles. As such, the main as well as the minor fatty acids are very useful in distinguishing among these strains [13].

Different techniques and methods for fatty acid analysis have been developed since Abel et al. (1963) [14] proposed quantitative and qualitative analyses for characterizing microorganisms, where it was indicated that gas chromatography (GC) has the necessary sensitivity, rapidity and selectivity for such analyses. They considered that this is an extremely selective method, such that a single normal-sized colony of microorganisms is sufficient. This method requires a short time between preparation to examination, and organic and inorganic nutrients do not interfere. The fatty acid methyl esters (FAMEs) from cell membranes have been analysed using GC. These fatty acids were extracted from cell hydrolysates and derivatized to volatile methyl esters for further detection by GC [14].

Other detection methods for fatty acid analysis have also been applied. These include gas-liquid chromatography and mass spectrometry (MS), which is a rapid, automated method for analysing mixed populations, and the various configurations of MS offer combined advantages of speed, sensitivity and selectivity. The method also requires that molecules be vaporized into the gas phase before they can be ionized and analysed.

Gas chromatographic or high pressure liquid chromatographic separation increases the capability for analysing complex mixtures, and they together provide two complementary kinds of characterization. Such chromatographic separation could even distinguish chirality, which is usually invisible to MS [15]. Fatty acids analysis and quantification by GC is carried out by the extraction of lipids from microorganisms, followed by hydrolysis and methylation. The resulting FAME profiles are used for the identification and differentiation of these microorganisms [16]. Actually, there are standardized and automated techniques to asses these compounds; using the Microbial Identification System (MIS) it is a possibility, in a short time, for acquiring precise information about bacteria fatty acid compositions. This system is applied most frequently in the clinical market and it is limited in the number or environmental species they can identify. Nevertheless, microbial fatty acid profiles are unique from one species to another [2].

The MIS system – also known as the MIDI Sherlock Holmes System – which identifies microorganisms using FAMEs and GC, consists of large microbial libraries; the procedure for FAME analysis [17] may be summarized by the following steps: 1) harvesting (the removal cells from the culture media), 2) saponification (the lysis of the cells to liberate fatty acids), 3) derivatization (methylation, the formation of FAMEs), 4) extraction (the transfer of the FAMEs from the aqueous phase to the organic phase by the use of an organic solvent), 5) base wash (an aqueous wash or the organic extract prior to chromatographic analysis).

Alternative procedures have also been proposed for the extraction and separation of FAMEs without the previously described steps, such as by the direct injection of the culture broth, allowing the separation and quantification of alcohols and volatile and non-volatile fatty acids. This method simplifies the procedure for fatty acid identification and quantification [18].

Pyrolysis-MS is another technique that allows the direct analysis of bacteria (depending on the organism). With pyrolysis, the bacteria are broken (in an inert atmosphere) using heat to produce volatile, low molecular weight substances that are detected and quantified. This technique involves the analysis of the total cellular composition of the bacteria, and a single bacterial colony is sufficient for analysis [19].

This technique can be used to provide GC data from non-volatile/solid materials. A GC with an ion trap mass spectrometer (IT-MS) could be used for this kind of analysis [20].

2.1. Fatty acid definition

Lipids are defined as natural products that may be isolated from biological materials by extraction with organic solvents and which are usually insoluble in water. This definition includes sterols, terpenoids, isoprenoids, waxes and pigments, many of which are not present in bacteria [21], and the simplest are the fatty acids, as in Table 1. These are the fundamental building blocks of complex lipids, like fatty acyls, glycerolipids, glycerophospholipids, sphingolipids, sterol lipids, prenol lipids, saccharolipids and polyketides [22].

Microbial fatty acids are typically 12-24 carbons long and the common membrane fatty acids are 14-20 carbons long. These compounds are either ester- or ether-linked. The latter is rare, and has been found in the phospholipids of Archaea [23]. In general, bacteria contain fatty acids between C10 and C20 in length, but there are some eubacteria species that have polyunsaturated fatty acids, like *Shewanella gelidimarina, Shewanella hanedai, Colwellia psychrerythraea* [24], *Shewanella donghaensis* [25], cyanobacteria and mycobacteria.

Fatty acids are essential to cells and their metabolic functions. These compounds are typically associated with energy storage and the structural fluidity of membranes; they form compounds that play important roles in cell signalling processes [26]. In addition, they function as metabolic intermediates and form part of the higher molecular weight lipids [27]. These carbon-based 'backbone' chains, may be elongated, shortened or altered by the activity of the elongase or desaturase enzymic activity, respectively, and longer fatty acids are results of lipogenesis. Moreover, the size of the fatty acid structure depends upon the addition or removal of double hydrogen bonds and catabolic activity [26].

Structurally, fatty acids are long chain monocarboxylic acids of the general formula R.COOH. Usually, the R-group has a long chain - commonly unbranched - consisting of an even number between eight and 24 carbon atoms [27]. The chain has a hydrophobic aliphatic tail that is either saturated or unsaturated [22], depending on the presence of double bounds in the R-group. Acids with one double bound are known as 'mono-enoic', those with two as 'di-enoic', and so on [27].

Fatty acids are structurally diverse and contain distinct classes that are defined at the molecular level, based on the degree of branching, the number and position of double bonds, the chain length, the *cis-trans* isomer conformations, and the presence of functional groups [22].

The major groups of fatty acids are: a) straight chain, b) branched chain, c) unsaturated, and d) cyclopropane. Straight chain fatty acids are found elsewhere in nature (e.g., C12, C14, C16, C18 and C20). However, the corresponding β– hydroxymyristic acid 3-OH and C12 are not normally found in higher organisms, but are common components of the boundary of lipids present in the walls of gram-negative bacteria. Branched fatty acids comprise two types: the iso-form (the methyl group is located on the penultimate carbon atom) and the anteiso (the methyl group is located in the antepenultimate carbon atom). The unsaturated acids found in bacteria are monounsaturated, whereby the double bound is most frequently found between carbon 11 and 12. The cyclopropane fatty acids - rarely found in higher organisms - are frequently encountered in bacteria. The biosynthetic precursors of these acids are the corresponding monounsaturated fatty acids; as a consequence, the major example is the cis-11-12 methylene hexadecanoid acid [21]. Table 1 contains some examples of fatty acids commonly found in bacteria.

Fatty acid	Simplified formula	Common name	Systematic name
SATURATED			
$C_{10}H_{20}O_2$	C16:0	Capric	Decanoic
$C_{12}H_{24}O_2$	C12:0	Lauric	Dodecanoic
$C_{14}H_{28}O_2$	C14:0	Myristic	Tetradecanoic
$C_{16}H_{32}O_2$	C16:0	Palmitic	Hexadecanoic
$C_{17}H_{34}O_2$	C17:0	Margaric	Heptadecanoic
$C_{18}H_{36}O_2$	C18:0	Stearic	Octadecanoic
$C_{20}H_{40}O_2$	C20:0	Arachidic	Eicosadecanoic
UNSATURATED			
$C_{16}H_{30}O_2$	C16:1ω9	Palmitoleic	Hexadec-9-enoic
$C_{18}H_{34}O_2$	C18:1ω9	Oleic	Octadec-9-enoic

Table 1. Examples of fatty acids found in bacteria: formula, common and systematic names [modified from 27]

2.2. Nomenclature of fatty acids

Fatty acids are designated by the total number of carbon atoms followed by the total number of double bonds (e.g., a 16 carbon alkanoic acid is 16:0), beginning with the position of the double bound closest to the methyl-end (ω) of the molecule. The double bond position is indicated by the number closest to the carboxyl-end of the fatty acid molecule with the geometry of either c (*cis*) or t (*trans*). For example, 16:1ω9*c* is a phospholipid fatty acid with a total of 16 carbons and with one double bond located between nine and 10 carbons from the methyl-end of the molecule in the *cis* configuration [45].

2.3. Analysis of fatty acids from photosynthetic bacteria

Numerous papers have been published about the influence of the culture media in growing bacterial cells as well as the salinity in the bacteria fatty acid compositions. These have been done especially in heterotrophic bacteria, and it has been observed that when *Escherichia coli* was grown in a limited concentration of ammonium salts, these bacteria produced high quantities of cyclopropane fatty acids [28].

The analysis of salinity effects on the fatty acid composition of photosynthetic bacteria has been done in PPSB, such as *Chromatium purpuratum* (or *Halochromatium purpuratum*), *Ectothiorhodospira mobilis* and *Ec. halophila* [29], using different salt concentrations; the results showed only quantitative alterations, but their fatty acid pattern did not witness changes.

The age of the cells is another factor that may change the fatty acid composition of bacteria as well as the cells of all living organisms. In many techniques, it has been suggested that it is very important to use bacterial cells of the same age for fatty acid analysis, especially for chemotaxonomical purposes.

In order to study the influence of $MgSO_4$ and NaCl and the age of the cells in the fatty acid composition of GPSB and PPSB, respectively, the bacterial strains included in Table 2 were used. This table contains the origin, the phenotypical characteristics and optimal culture conditions (NaCl) of the microorganisms.

	Strain	Origin	Main pigments	Shape	NaCl(%)*
Chromatiaceae					
T9s642	[a]*Chromatium* sp.	TL, Mexico	Bchl *a, spx*	Rods	2
DSMZ4395	[a]*Chr. salexigens*	GSC, France	Bchl *a, spx, ly, rh*	Rods	7
Chlorobiaceae					
T11s	[a]*Chlorobium* sp.	TL, Mexico	Bchl *c, cb*	Rods	2
DSMZ249	*Cb. limicola f.***	THS, USA	Bchl *c, cb*	Rods	0
DSMZ266	*Cb. phaeobacteroides*	LB, Norway	Bchl *e, irt*	Rods	0

DSMZ (=DSM) = Deutsche Sammlung von Mikroorganismen und Zellkulturen GmbH; GSC = Giraud Saltens Camargue; TL = Tampamachoco Lagoon (Veracruz); THS = Tassajara Hot Spring; LB = Lake Blankvann; *Bchl* bacteriochlorophyll, *spx* spirilloxanthin, *ly* lycopene, *cb* chlorobactene, *rh* rhodopinal; *NaCl (%) = optimal salinity requirement.

[a]Marine origin strains. The genus *Chromatium* was reclassified into *Halochromatium*. ***Chlorobium limicola f. thiosulfatophilum*=actually *Chlorobaculum thiosulfatifilum*

Table 2. List of strains used in this work: their origins and optimal salinity

For all the experiment, two wild strains were used: a PPSB strain T9s642 (*Chromatium* sp.) and a GPSB strain T11s (*Chlorobium* sp.). These photosynthetic bacteria were isolated from superficial water samples collected in Tampamachoco, a coastal lagoon located in Tuxpan (Veracruz, Mexico) [7].

For culturing these bacteria, a basal medium enriched with water samples was used. This was also used for the isolation and growth of these photosynthetic bacteria and contained the following: distilled water 950 ml, KH_2PO_4 1.0 g, NH_4Cl 0.50 g, $MgSO_4 \cdot 7H_2O$ 0.40 g, $CaCl_2 \cdot 2H_2O$ 0.50 g, and NaCl 2.0% (only for marine strains). It was sterilized by autoclave and, once cooled, was supplemented with a 1.0 mL SL10 solution (GPSB) and SL12 (PPSB), Vitamin B12 1.0 mL

(2.0 mg/100 mL distilled water), and 30 mL of sodium bicarbonate solution 5% [30]. 6.0 mL and 10.0 mL of $Na_2S \cdot 9H_2O$ 5% (v/v distilled water) were added as an electron donor for *Chromatium* and *Chlorobium*, respectively. The pH was adjusted to 7.5 for the culture medium of *Chromatium* and 6.5 for *Chlorobium*.

2.3.1. *Influence of $MgSO_4$ and NaCl on cultures of GPSB*

The influence of $MgSO_4$ and NaCl on cultures of GPSB (T11s) were observed by growing the bacteria in the basal medium, as described before, and modified as follows: condition T11S = NaCl (2.0%) and $MgSO_4 \cdot 7H_2O$ 0.3% g; condition C11S = NaCl (0.0%), $MgSO_4 \cdot 7H_2O$ 0.04% g; and condition MG11S = NaCl (0.0%) and $MgSO_4 \cdot 7H_2O$ 0.3% g, respectively. Conditions T11S and C11S are optimal for marine and freshwater GPSB, respectively. In all cases, the pH was adjusted to 6.5 using a diluted H_2S sterilized solution. Strains of GPSB, DSMZ249 and DSMZ 266 were used as references, and they were cultured under the condition C11S (optimal for freshwater strains).

2.3.2. *Influence of cell age on the fatty acid composition of PPSB*

In order to investigate the influence of bacterial cell age on the fatty acid composition of PPSB, a wild strain of *Chromatium* sp. (*Halochromatium* sp.), isolated from Tampamachoco lagoon, and a collection strain of *Chromatium salexigens* DSMZ4395, were used. The fatty acids of the cultures of these bacteria were analysed after 15, 20 and 30 days of growth, and they were incubated under the standard conditions as described previously.

All the bacteria cultures (green and purple) were grown in 1.5 L glass flasks, the temperature of incubation was 23ºC and, as the light source, bulbs of incandescent (for *Chromatium* strains) and fluorescent light (*Chlorobium* strains) were used. In all cases, the light intensity was 2000 Lux.

The mass cultures of GPSB and PPSB were obtained by adding a previously sterilized and neutralized solution of $Na_2S \cdot 9H_2O$. It was added to the cultures when the sulphur source (electron donor) was depleted, which was detected using an acetate paper pH tester. Cultures were grown for 15 days (with the exception of *Chromatium* cultures, as indicated before) under the conditions described. Once the absence of sulphur in the bacterial cultures was detected, they were centrifuged (5,000-1000 rpm) and afterwards freeze-dried for further methylation and fatty acid analysis.

2.3.3. *Fatty acid extraction and analysis by GC and GC-MS*

The procedure for total fatty acid analysis has been described previously in [7], and the lyophilized cells of GPSB and PPSB (250 mg and 500 mg, respectively) were used for fatty acid extraction by saponification followed by diazomethane derivation. The resulting FAMEs were identified using a standard methyl ester mixture and the final identification was confirmed using GC and MS (GC-MS). Cell lipids from each sample were saponified by adding 2.0 mL benzene and 8.0 mL KOH solution (5.0 g/100 mL methanol), containing 5% (w/v) under a temperature of 80°C (using a cover bath), over four hours. Once the sample was cooled to room temperature, it was acidified to pH = 1 with a H_2SO_4 solution (20% v/v).

FAMEs were prepared by adding to each sample 2.0 g N-Nitroso-N-methylurea (Sigma) dissolved in a pre-cooled solution of 30.0 mL diethyl ether, 2.0 g KOH and 6.0 mL distilled water. This mixture was stirred for 5-10 minutes and the supernatant was removed and placed in a new tube containing KOH pellets cooled on ice. Finally, diazomethane was added and the dried lipids were achieved within 10-20 min and the content was evaporated at 40°C in a water bath.

The FAMEs were analysed by injecting 2.0 μL of the sample, previously dissolved with n-hexane (0.04-1.0 mL), in a HP-5890A gas chromatograph equipped with a flame ionization detector and a silica capillary column (15 m x 0.25 mm I.D.) with cross-linked methyl silicone (HP-1, Hewlett-Packard). The column was programmed from 175°C to 300°C within 15 minutes. The injector and detector temperature were 275°C and 300°C, respectively. Helium was used as a carrier gas with a flow rate of approximately 1.0 mL/min, and the split ratio was approximately 1:50. Finally, a HP3396 integrator was used for the chromatogram integration and the identification of the FAMEs was done by comparing the retention time of each fatty acid, using a standard mixture of FAMEs.

For the GC-MS analysis, an HP-5890 gas chromatograph attached to a HP5989X quadripole mass spectrometer was used with a methyl polysiloxane column TRB1 (30 m). The injector and detector temperatures were both 225°C, and two ramps were used: one from the initial temperature 10°C/min to 240°C and the other with 40°C/min to 270°C. The injection mode was splitless.

The standard deviations and means were calculated based on the integration areas of the fatty acid peaks.

3. Results

3.1. Influence of the $MgSO_4$ and NaCl cultures on GPSB

Strain T11S was isolated from superficial water samples from Tampamahoco lagoon (Veracruz, México). One of the characteristics of this bacterium is that it is able to tolerate physical and chemical environmental changes. For this reason, it was selected to investigate the changes in fatty acid composition due to salinity (NaCl and $MgSO_4$). However, no changes of the fatty acid profile have been observed as a response to NaCl and $MgSO{\cdot}7H_2O$ changes. In all the strains analysed, 12 fatty acids comprising saturated, unsaturated, branched and cyclic acids were detected, as has been reported previously [7].

All the fatty acids detected in the GPSB had chains with no more than 18 carbons (Table 2). In the *Chlorobium* collection strains (DSMZ266, DSMZ249) and in T11S, the same fatty acids were detected. The standard deviations for almost all the fatty acids identified were very small, with the exception for the C18:0 fatty acid.

Wild and reference strains of *Chlorobium* have C14:0 (6.14-21.3%), C16:0 (11.9-22.9%) and C16:1 (26.9-31%) as the main fatty acids, a profile that was reported for the same strains, cultured in similar conditions [7]. These results confirm the use of FAMEs to characterize bacteria, because they are conserved in composition and proportion if they are grown in similar conditions [11].

All the strains were incubated in similar physical (temperature and illumination) conditions, but specific culture conditions were used for each one. The culture medium for DSMZ266 and DSMZ249 was the same as with C11S (this is without NaCl and 0.04 g $MgSO_4 \cdot 7H_2O$), but it is evident that the condition C11S conserves the same fatty acid pattern as T11S and MG11S.

The collection strains have less saturated fatty acids than T11S, while MG11S and C11S have similar quantities of these fatty acids; inversely, the collection strains have more unsaturated fatty acids than the experimental cultures (T11S, MG11S and C11S). It is very important to note that the branched fatty acids recorded were similar to all the strains cultured in the same conditions (DSMZ266, DSMZ249 and C11S); it is evident that the bacterium of marine origin (T11S) cultured in optimal conditions produced more branched fatty acids than the strains cultured in freshwater conditions as in DSMZ266 and DSMZ249. It has been pointed out that the presence of branched chain fatty acids confers to the membrane a greater degree of flexibility [31].

The cyclic fatty acid was present in greater quantities in DSMZ249 than in DSMZ266 and T11S. When there was an increase in NaCl or Mg salts, the production of C17Cy fatty acid decreased too. It is evident that in optimal conditions for T11S, there was less production of C17Cy fatty acid, so the presence of this fatty acid is as follows: MG11S>C11S>T11S. Cyclopropane acid production in cells is a mechanism for preserving membrane integrity, preventing the formation of abnormal structures [32].

The values of the standard deviation for each fatty acid of T11S, MG11S and C11S were between 0.009 and 1.828, which is evidence that the changes in the fatty acid compositions of *Chlorobium* sp. due to salinity (NaCl and $MgSO_4$) were not so large to loss its identity (Table 3).

3.2. Influence of the culture age (20 and 30 days of growth) on the fatty acids composition of PPSB

It has been proposed that cell age is among the factors that affect the fatty acid compositions of living organisms. The influence of this factor in the fatty acid composition of bacterial cells has been studied in heterotrophic bacteria, and cultures of 24 and 48 hours incubation have been proposed as the optimal age (Microbial Identification System) for the fatty acid analysis of these bacteria. However, the exponential phase has been proposed for fatty acid analysis in photosynthetic bacteria [33]. In addition, stationary or low stationary phases, after three and six days of growth, have been indicated too [34]. Nevertheless, and frequently, the age of the cells that they used for fatty acid analysis was not mentioned.

In the present experiment, the cells of two strains of PPSB from marine origin were used. For fatty acid analysis, the cells of these bacteria were collected after 15, 20 and 30 days of growth in optimal conditions according to their origins. They were also incubated under the same light and temperature conditions. The results of the fatty acid analysis of *Chromatium salexigens*

Fatty acids	DSMZ249 (3)		DSMZ266 (2)		T11S (4)		MG11S (4)		C11S (3)	
SATURATED	X	SD	X	SD	X	SD	X	SD	X	SD
C12:0	1.489	0.342	0.970	0.414	1.943	0.156	1.881	0.910	2.384	0.264
C14:0	6.114	2.353	13.626	0.406	20.606	1.001	17.366	1.371	21.305	0.860
C15:0	0.757	0.076	1.119	0.532	0.191	0.093	0.441	0.096	0.278	0.162
C16:0	11.932	0.621	11.960	0.622	22.973	1.828	21.435	0.492	19.949	0.696
C17:0	0.230	0.158	0.246	0.050	0.202	0.035	0.429	0.009	0.224	0.015
C18:0	0.838	0.066	0.663	0.252	0.660	0.255	0.334	0.364	0.325	0.018
Sum (%)	21.360		28.584		46.574		41.888		44.465	
UNSATURATED										
C16:1(9)	31.034	0.293	36.009	0.009	26.912	1.172	26.912	0.922	27.215	1.524
C18:1	10.298	0.055	2.689	2.147	2.972	0.188	3.255	0.107	4.643	1.754
Sum (%)	41.332		38.697		29.884		30.167		31.858	
BRANCHED										
12-Me-C14:0	0.516	0.016	0.421	0.070	0.458	0.296	0.973	0.070	0.272	0.013
5-Me-C16:0	0.632	0.078	0.538	0.100	1.163	0.041	1.477	0.335	1.326	0.076
aC17:0	0.862	0.097	0.608	0.011	1.383	0.053	0.770	0.042	0.595	1.123
15-Me-C16:1	0.620	0.176	0.461	0.061	ND	-	ND	-	0.239*	-
Sum (%)	2.629		2.029		3.004		3.220		2.432	
CYCLIC										
17Cy	3.336	0.146	1.780	0.271	0.827	0.038	2.014	1.897	1.426	0.051
TFAE (%)	68.657		71.090		81.289		77.288		80.181	

TFAE = Total fatty acid extracted; X = mean (%); SD = Standard deviation; Experimental culture (number of independent cultures analysed) ND = not detected.

*It was detected only in one of the three cultures of the same analysed strain.

Table 3. Influence of NaCl and $MgSO_4 7\ H_2O$ on the fatty acid composition of a wild strain of GPSB (T11s)

DSM4395 (*Halochromatium salexigens*) and *Chromatium* sp. T9s642 are shown in Table 4 and Figures 1 and 2, respectively.

The fatty acid analysis of these bacteria revealed the presence of nine of these compounds. All these fatty acids have been reported before for members of the Chromatiaceae family [7, 9]. The major fatty acids were C18:1, C16:1 and C16:0 and less quantities of: C12:0, C14:0 and C17:0, C18:0 and C20:1. These conformed between 92% and 93% of the total compounds extracted from the cells of two *Chromatium* strains (DSMZ4395 and T9s642). In this analysis, fatty acids less than 0.1% were also taken into account because minor fatty acids are very important in identifying microorganisms, as reported by Núñez-Cardona [7, 8]. For some strains of Cyanobacteria,it has been reported trace quantities of fatty acids (less than 0.03%) [8], and it has been indicated that minor fatty acids could be useful as biomarkers.

There were no qualitative changes in the fatty acid compositions of DSMZ4395 and T9s642 due to cell ages, and the quantitative changes were minimal: both strains conserved the fatty acid pattern that it is characteristic for members of the Chromatiaceae family, but strain DSMZ4395 produced almost the same quantities of C16:0 (16.2-16.8%) and C16:1 (17.5-18.2%), which is characteristic of halophilic microorganisms like *Halochromatium purpuratum* [29] and

	DSMZ4395-30		DSMZ4395-15		T9s642-20		T9s642-15	
Fatty acid	X	SD	X	SD	X	SD	X	SD
C12:0	0.680	0.015	0.718	0.129	1.049	0.253	0.705	0.057
C14:0	0.873	0.064	0.837	0.054	0.201	0.010	0.237	0.008
C16:1	18.283	0.387	17.520	0.527	27.714	0.102	27.414	0.336
C16:0	16.195	1.249	16.810	0.200	10.498	0.327	11.605	0.887
C17:0	0.428	0.085	0.330	0.069	0.289	0.132	0.413	0.064
C18:1(9)	48.911	1.193	50.526	2.405	44.674	0.141	45.893	0.219
C18:1(11)	5.005	0.223	4.773	0.085	5.757	0.023	5.622	0.296
C18:0	2.784	0.260	2.696	0.272	2.168	0.444	1.928	0.145
C20:1	0.624	0.030	0.724	0.084	0.634	0.004	0.627	0.011
TFAE (%)	93.783		94.934		92.983		94.445	

Table 4. Average (%) and standard deviation (SD) of each fatty acid identified in whole cells of *Chromatium* strains, concentrated at 15, 20 and 30 days of growth

Thiocapsa halophila [7]. Nevertheless, in *Thiohalospira halophila,* a chemolithoautotrophic, halophilic, sulphur-oxidizing gammaproteobacteria member of *Ectothiorhodospiraceae,* fatty acids such as 10MeC16:0, C16:0 and C18:0 are predominant (more than 80%). Furthermore, it has been pointed out that C16:0 is often the main fatty acid in halophilic bacteria [35]. However, this was not applicable for DSMZ4395. In this strain, the major fatty acids are the unsaturated acids (C16:1 and C18:1). This property was characteristic of other halophilic chemolithoautotrophic bacteria, like *Thiohalomonas denitrificans* (DSM15841), where C18:1 is less than 18.58% (Σ:C18:1ω9, 18:1ω7, C18:1ω5) [36]. In contrast, in the *Chromatium salexigens* strain (DSMZ4395) this is the main fatty acid (48.9-50.5%).

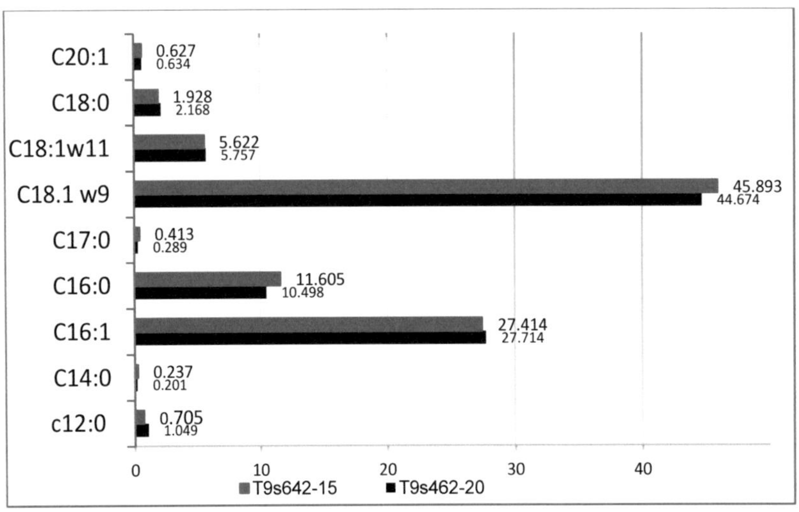

Figure 1. Fatty acids (%) of the wild culture T9s642-15 and T9s642-20 from cells centrifuged after 15 and 20 days of growth, respectively.

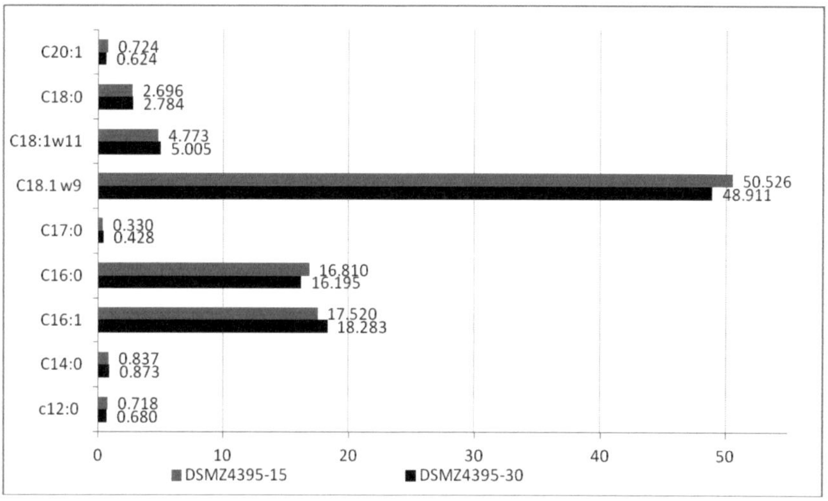

Figure 2. Fatty acids (%) of DSMZ4395-15 and DSMZ4395-30, obtained in cells centrifuged after 15 and 30 days of growth under standard conditions, respectively.

On the other hand, in the wild *Chromatium* sp. strain T9s642, as in some mesophilic eubacteria, the fatty acid C16:1 was present in greater quantities than C16:0 and *Chromatium salexigens* (DSMZ4395) produced more saturated fatty acids (Figures 1 and 2). Nevertheless, it is evident that for the fatty acid compositions in both bacteria, the age factor causes a increase of C18:1 and C16:0 in younger cells, and also a small increase of C16:1, in the oldest.

The influence of age on the fatty acid compositions of bacteria has also been studied in microorganisms like *Serratia marcescens* and *Escherichia coli*, which have more C16:0, C16:1 and C18:1 as well as more cyclic fatty acids. When these cultures are in stationary phases (24-48 hours), monounsaturated fatty acids (e.g., C16:1 and C18:1) are transformed to cyclic forms (17Cy and 19Cy) and C16:0 increases too [37].

The changes in branched fatty acids, such as iC15:0 and C15:0 (e. g , increases and decreases, respectively), are the result of aged cells in a strain of *Bacillus* [38]. This microorganism, as in many other gram-positive bacteria, is characterized by high quantities of branched fatty acids, which are responsible to maintain the fluidity and the membrane structure.

3.3. Uses of fatty acids analysis (other than chemotaxonomy)

Two goals of chemotaxonomy are to measure the number of changed molecules as a descriptor of the chemical relevance of two bacteria and also to provide a list of changed molecules to discover biomarkers [39].

Bacteria contain lipids in concentrations of 0.2-50% of dry weight, and for fatty acid analysis lipid compositions are important. These are mainly: phospholipids (structural elements of the cell membrane), glycolipids (elements of cell membrane structures but less common than phospholipids; abundant in Actinomycetes), lipid A (present in the outer membrane of gram-negative bacteria; part of the lipopolysaccharides) and lipoteichoic acids (present in gram-positive bacteria) [2].

Profiles of fatty acids have been used to distinguish pathogenic bacteria species for fish, like the genus *Tenacibaculum* (agents of flexibacteriosis in marine fish) whose cells are characterized by the presence of large amounts of branched (36.1–40.2%) and hydroxylated (29.6–31.7%) fatty acids [40].

The presence of bacteria in sponges has been detected because they produce large quantities of iso-, anteiso-cyclopropyl- and monomethyl-branched fatty acids. Using fatty acid analysis, it is possible to distinguish different symbionts (photosynthetic and non-photosynthetic) present in marine sponges and provide estimates of bacterial symbiont abundances [41].

Special fatty acids from phospholipids have been used for characterizing bacteria and photosynthetic life forms. Fatty acids linked to these molecules in the members of Chromatiaceae and Ectothiorhodospiraceae were analysed, and it is evident that the fatty acid pattern does not change. As has been indicated before, these compounds have great value in determining bacterial phylogeny, providing a useful set of features for characterizing strains, and they give important information about microbial communities in environmental samples.

Phospholipid fatty acids could be used as biomarkers because they degrade rapidly after cell death and are not present in storage lipids or in anthropogenic contaminants. Bacteria contain phospholipids as relatively constant proportions in their biomass [41], and with these it is possible to estimate the microbial biomass in environmental samples where the bacteria of subsets in microbial communities contain specific signature fatty acids in their phospholipids [42]. Using phospholipid fatty acids (as methyl esters) analysis, it is possible to measure attributes or microbial communities, such viable biomass and microbial structures as well as nutritional and physiological status [41, 43]; these compounds, as well as bacterial fatty acid ratios, have been used to disclose bacterial connections in the marine food web and their importance in supplying material and energy to the higher trophic levels [44].

In addition, lipids have been used to establish possible links between modern and ancient microbial communities and to add to the understanding of the evolutionary histories of the organisms in the various environments. The analysis of phospholipids provides information about the viable biomass and biological diversity, and on the other hand information about the glycolipids of the activity of photosynthetic microorganisms in the environment. In addition, sterols and hopanoids - which are better preserved in the geological records - offer insights into the presence of microorganisms in paleoenvironments [45]. In bacterial membranes, the fatty acids of phospholipids are linked by ester bonds to glycerol and form a phospholipid bilayer, while archaeal membranes have phospholipids containing two polar heads linked by isoprenoid chains and ether linkages to glycerol [44]. As such, these linkages are markers for detecting archaeal bacteria.

3.4. Lipidomics studies: some examples

Lipidomics has been applied in physiological studies, such as the change in the lipidomic profiling of *Chlamydomonas nivalis* as a response to nutritional stress [46]. In *Nitzschia closterium*

http://dx.doi.org/10.5772/57389

f. minutissima, lipid metabolomic changes in the different growth phases of these microorganisms have been studied. Some biomarkers have been selected and identified, including free fatty acids and lipids such as harderoporphyrin, phosphatidylglycerol, 1,2-diacylglycerol-3-O-40-(N,N-trimethyl)-homoserine, triacylglycerol, cholesterol, sulfoquinovosyldiacylglycerol, lyso-sulfoquinovosyldiacylglycerol, monogalactosyldiacylglycerol, digalactosyldiacylglycerol and lyso-digalactosyldiacylglycerol [47].

Lipidomics has revealed that under cold stress, *Stephanodiscus* sp. produces the following compounds as biomarkers: triacylglycerol, phosphatidylcholine, phosphatidylglycerol, sulfoquinovosyldiacylglycerol, monogalactosyldiacylglycerol, lyso-phosphatidylglycerol, lyso-phosphatidylcholine, lyso-monogalactosyldiacylglycerol and lyso-sulfoquinovosyldiacylglycerol [48].

In addition, lipidomics have been applied to the understanding of the lipogenesis of free fatty acids from the symbiosis between cnidarian-dinoflagellate [26] as well as the study of the lipidomics of human pathogenic mycobacteria. It is possible to get the lipidomics datasets to assess lipid changes during infection or else among clinical strains of mycobacteria tuberculosis [39].

4. Conclusions

As was mentioned before, the influence of NaCl in the culture medium increases the concentration of the saturated fatty acids C16:0 and decreases the production of the unsaturated C16:1 and C18:1. The presence in the culture medium of both $MgSO_4$ and NaCl reduces the production of the saturated fatty acids C14:0 and C18:0 and increases the production of the unsaturated C16:1 and C18:1. The change of the relative abundances of the fatty acids due to the changes of $MgSO_4$ and NaCl in the culture media of *Chlorobium* is not sufficient to alter the fatty acid pattern of the *Chlorobium* strains analysed. Minimal quantitative alterations were observed by the cell age, the pattern and profile of the fatty acids in PPSB and GPSB are very conservative in these microorganisms could be used for chemotaxonomic approaches. Nevertheless it is very important to describe the culture conditions as well as the techniques for fatty acid analysis.

Although advances in the technologies for fatty acid analysis have been reported, the isolation and culturing of novel strains as well as the fatty acid analysis of photosynthetic prokaryotes remains scarce, and it is very important to encourage the investigation of these microorganisms. With new sources of genetic materials from photosynthetic bacteria and the application of new technologies - for example lipidomics - which are more precise, it will be possible to discover new fatty acid structures and to evaluate the sources of these compounds of biotechnological interest and novel chemotaxonomic biomarkers. In addition, the physiological role of fatty acids in the cell membranes of photosynthetic bacteria and their role in environmental trophic chains could be explored.

Acknowledgements

We thank the Universitat Autónoma de Barcelona (UAB), Spain, where all the experimental elements of the present work were performed with academic and technical support from Doctors Marina Luquin, Jodi Mas and Manuel Muñoz. Financial support from the Consejo Nacional de Ciencia y Tecnología (CONACyT, México) and the Universidad Autónoma Metropolitana-Xochimilco are gratefully acknowledged.

Author details

María Teresa Núñez-Cardona

Universidad Autónoma Metropolitana-Xochimilco. Calzada del Hueso, México, Distrito Federal, Mexico

References

[1] Gupta RS, Mukhtar T, and Singh B. Evolutionary relationships among photosynthetic prokaryotes (*Heliobacterium chlorum*, *Chloroflexus aurantiacus*, Cyanobacteria, *Chlorobium tepidum* and Proteobacteria): implications regarding the origin of Photosynthesis. Molecular Microbiology 1999; 32(5) 893-906. DOI: 10.1046/j.1365-2958.1999.01417.x.

[2] Janse JD, 1992 Whole cell fatty acid analysis as a tool for classification of phytopathogenic pseudomonas bacteria PhD Thesis. p. 121 http://Library.wur.nl/WebQuery/clc/338144 accessed in 08 08 2013

[3] Imhoff JF, Caumette P. Recommended standards for the description of new species of anoxygenic phototrophic bacteria. International Journal Systematic and Evolutionary Microbiology 2004; 54(4) 1415-1421. DOI: 10.1099/ijs.0.03002-0.

[4] Tindall BJ, Rosselló-Móra R, Busse HJ, Ludwig W, and Kampfer P. Notes on the characterization of prokaryote strains for taxonomic purposes. International Journal of Systematic and Evolutionary Microbiology 2010; 60(1) 249–266. DOI: 10.1099/ijs.0.016949-0.

[5] Knudsen E, Jantzen E, Bryan K, Ormerod JG, and Sirevag R. Quantitative and structural characteristics of lipids in *Chlorobium* and *Chloroflexus*. Archives of Microbiology 1982; 132(2) 149-154. http://link.springer.com/article/10.1007/BF00508721.

[6] Kenyon CN, Gray AM. Preliminary analysis of lipids and fatty acids of Green Bacteria and *Chloroflexus aurantiacus*. Archives of Microbiology 1974; 120(1) 131-178. http://www.ncbi.nlm.nih.gov/pmc/articles/PMC245741/pdf/jbacter00334-0147.pdf.

[7] Núñez-Cardona, MT (2012). Fatty acids analysis of photosynthetic sulfur bacteria by gas chromatography. Gas chromatography-Biochemicals, Narcotics and Essential Oils. In Tech Publications, 117-138. http://cdn.intechopen.com/pdfs/31528/InTech-Fatty_acids_analysis_of_photosynthetic_sulfur_bacteria_by_gas_chromatography.pdf.

[8] Caudales R, Wells J. Differentiation of Free-Living *Anabaena* and *Nostoc* Cyanobacteria on the basis of fatty acid composition. International Journal of Systematic Bacteriology 1992; 42(2) 246-251. DOI: 10.1099/00207713-42-2-246.

[9] Núñez-Cardona MT, Donato-Rondon CCh, Reynolds CS, Mas J. A purple sulfur bacterium from a high-altitude lake in the Colombian Andes." Journal of Biological Research. Thessalon 9 2008; 17-24. http://www.jbr.gr.

[10] Bryantseva IA, Tourova TP, Kovaleva OL, Kostrikina NA, and Gorlenko VM. *Ectothiorhodospira magna* sp. nov., a new large alkaliphilic purple sulfur bacterium. Microbiology 2010; 79(6) 780-790.

[11] Walcott R, Langston D, Sanders F, and Gitaitis R. Investigating Intraspecific Variation of *Acidovorax avenae* subsp. *citrulli* using DNA fingerprinting and whole cell fatty acid analysis. Phytopathology 2000; 90(2) 191-196. http://dx.doi.org/10.1094/PHYTO.2000.90.2.191.

[12] Tighe SW, De Lajudie P, Dipietro K, Lindström K, Nick G, and Jarvis B. Analysis of cellular fatty acids and phenotypic relationships of *Agrobacterium, Bradyrhizobium, Mesorhizobium, Rhizobium* and *Sinorhizobium* species using the Sherlock Microbial Identification System. International Journal of Systematic and Evolutionary Microbiology 2000; 50(2) 787–801. DOI: 10.1099/00207713-50-2-787.

[13] Moss C, Lambert M, and Merwin W. Comparison of Rapid Methods for Analysis of Bacterial Fatty Acids. Applied Microbiology 1974; 28(1) 180-185. http://www.ncbi.nlm.nih.gov/pmc/articles/PMC186596/pdf/applmicro00013-0102.pdfhttp://aem.asm.org/content/28/1/80.short.

[14] De Gelder J. Raman spectroscopy as a tool for studying bacterial cell compounds PhD in Sciences Chemistry Universiteit Gent. 2008. http://hdl.handle.net/1854/LU-472114. Accessed in 03 08 2013.

[15] Fenselau C. Mass spectrometry for characterization of microorganisms an overview. ACS Symposium Series. American Chemical Society: Washington DC. 1995 Chapter 1; 1-6 http://pubs.acs.org/doi/pdf/10.1021/bk-1994-0541.ch001.

[16] Bastida F, Kandeler E, Moreno JL. Ros M, Garcıa CA, and Hernandez T. Application of fresh and composted organic wastes modifies structure, size and activity of soil microbial community under semiarid climate. Journal of Analytical and Applied Pyrolysis 2005; 73 69–75. DOI:10.1016/j.apsoil.2008.05.007.

[17] Kunitsky, C., Osterhout G, and Sasser M. Identification of microorganisms using fatty acid methyl ester (FAME) analysis and the MIDI Sherlock Microbial Identification System. Encyclopedia of rapid microbiological methods 2006; 3: 1-18.

[18] Bories C, Rimbault A, and Leluan G. Simplified gas chromatographic procedure for identification and chemotaxonomy of anaerobic bacteria. Annals Institute Pasteur/ Microbiol. Microbiology 1987; 138, 587-592. http://www.sciencedirect.com/science/article/pii/0769260987900445.

[19] Brondz, I, Olsen, I. Microbial chemotaxonomy: chromatography, electrophoresis and relevant profiling techniques. Journal of Chromatography B: Biomedical Sciences and Applications 1986: 379 367-411. http://www.ncbi.nlm.nih.gov/pubmed/2426294.

[20] Akotoa L, Pelb R, Irtha H, Brinkmana U ATh., Rene, and Vreuls RJJ. Automated GC–MS analysis of raw biological samples Application to fatty acid profiling of aquatic micro-organisms. Journal of Analytical and Applied Pyrolysis. 2005; 73(1) 69–75. DOI:10.1016/j.jaap.2004.11.006.

[21] Shaw, N. Lipid Composition as a Guide to the Classification of Bacteria. Advances in Applied Microbiology 1974; 17: 63.

[22] Khalil BM, Hou W, Zhou H, Elisma F, Swayne LA, Blanchard AP, Yao Z, Bennett SAL, and Figeys D. Lipidomics era: Accomplishments and challenges. Mass Spectrometry Reviews 2010; 29(6) 877-929. DOI: 10.1002/mas.20294.

[23] Piotrowska-Seget Z, Mrozik A. Signature Lipid biomarker (SLM) analysis in determining changes in community structure of soil microorganisms. Polish Journal of Environmental Studies 2003 12(6); 669-675, http://www.pjoes.com/pdf/12.6/669-675.pdf.

[24] Russell N, Nichols D. Polyunsaturated fatty acids in marine bacteria–a dogma rewritten. Microbiology 1999; 145(4) 767-779. DOI: 10.1099/13500872-145-4-767.

[25] Yang S, Lee J-H, Ryu J-S, Kato Ch, Sang-Jin, Kim S-J. *Shewanella donghaensis* sp. nov., a psychrophilic, piezosensitive bacterium producing high levels of polyunsaturated fatty acid, isolated from deep-sea sediments. International Journal of Systematic and Evolutionary Microbiology 2007; 57 208–212. DOI: 10.1099/ijs.0.64469-0

[26] Dunn SR, Thomas MC, Nette GW, and Dove SG. A lipidomic approach to understanding free fatty acid lipogenesis derived from dissolved inorganic carbon within cnidarian-dinoflagellate symbiosis. PloS one 2012; 7(10) e46801. http://www.plosone.org/article/info%3Adoi%2F10.1371%2Fjournal.pone.0046801

[27] Hawker LE, Linton AH. Microorganisms function, form and environment 1971 pp. 727 P. London UK, Edward Arnold Ltd. ISBN 7131-2278-1.

[28] Marr AG, Ingrahan JL. Effect of temperature on the composition of fatty acids in *Escherichia coli*. Journal of Bacteriology 1962; 84 1262-1267. http://www.ncbi.nlm.nih.gov/pmc/articles/PMC278056/pdf/jbacter00464-0150.pdf.

[29] Imhoff, JF, Thiemann. B. Influence of salt concentration and temperature on the fatty acid compositions of *Ectothiorhodospira* and other halophilic phototrophic purple bacteria. Archives of Microbiology 1991; 156 (5) 370-375. http://link.springer.com/article/10.1007/BF00248713#page-1.

[30] Pfennig N, Trüper HG. Isolation of Members of the families Chromatiaceae and Chlorobiaceae. Springer Verlag 1981.

[31] Cho K, and Salton M. Fatty acid composition of bacterial membrane and wall lipids. Biochimica. Biophysica Acta 1966; 116(1) 73-79. http://www.sciencedirect.com/science/article/pii/0005276066900932.

[32] Valderrama M, Monteoliva-Sánchez M, Quesada E, and Ramos-Comenzana A. Influence of salt concentration of the cellular fatty acid composition of the moderately halophilic bacteria *Halomonas salina*. Research in Microbiology 1998; 149(9) 675-679. http://dx.doi.org/10.1016/S0923-2508(99)80015-1.

[33] Imhoff JF. Lipids, fatty acids and quinones in taxonomy and phylogeny of anoxygenic phototrophic bacteria, In Green Photosynthetic Bacteria. Plenum Press.1988 http://link.springer.com/chapter/10.1007%2F0-306-47954-0_10#page-1.

[34] Kenyon CN. Fatty acid composition of unicellular strains of blue-green algae. Journal of Bacteriology 1972; 109(2): 827-834 http://jb.asm.org/content/109/2/827.short.

[35] Sorokin DY, Tourova TP, Muyzer G, and Kuenen GJ. *Thiohalospira halophila* gen. nov., sp. nov. and *Thiohalospira alkalphila* sp. nov., novel obligately chemolithoautotrophic, halophilic, sulfur-oxidizing gammaproteobacteria from hypersaline habitats. International Journal of Systematic and Evolutionary Microbiology 2008; 58 1685–1692. DOI: 10.1099/ijs.0.65654-0.

[36] Sorokin DY, Tatjana P. Tourova, Gesche Braker, and Gerard Muyzer G. *Thiohalomonas denitrificans* gen. nov., sp. nov. and *Thiohalomonas nitratireducens* sp. nov., novel obligately chemolithoautotrophic, moderately halophilic, thiodenitrifying Gammaproteobacteria from hypersaline habitats. International Journal of Systematic and Evolutionary Microbiology 2007; 57: 1582–1589. DOI: 10.1099/ijs.0.65112-0.

[37] Law JH, Zalkin H, and Kaneshiro T. Transmethylation reactions in bacterial lipids. Biochimica et Biophysica Acta (BBA)-Specialized Section on Lipids and Related Subjects 1963; 70 143-151. http://www.sciencedirect.com/science/article/pii/0006300263907340.

[38] Hack SK, Garchow H, Odelson DA, Forney LJ, and Klug MJ. Accuracy, reproducibility, and interpretation of fatty acid methyl ester profiles of model bacterial communities. Applied and Environmental Microbiology 1994; 60(7) 2483-2493. http://pubmedcentralcanada.ca/pmcc/articles/PMC201674/pdf/aem00024-0287.pdf.

[39] Layre E, Sweet L, Hong S, Madigan CA, Desjardins D, Young DC, Cheng T-Y, Annand J-W, Kim K, Shamputa IC, McConnell MJ, Debono A, Behar SM, Minnaard AJ, Murray M, Barry CE, Matsunaga I, and Moody DB. Comparative Lipidomics Plat-

form for Chemotaxonomic Analysis of Mycobacterium tuberculosis. Chemistry and Biology 2011; 18 1537–1549. DOI 10.1016/j.chembiol.2011.10.013.

[40] Piñeiro-Vidal M, Pazos F, and Santo Y. Fatty acid analysis as a chemotaxonomic tool for taxonomic and epidemiological characterization of four fish pathogenic *Tenacibaculum* species Letters in Applied Microbiology 2008; (5): 548–554, DOI:10.1111/j.1472-765X.2008.02348.x.

[41] Guillan FT, Soilov IL, Thompson JE, Hogg RW, Wilkinson CR, and Djerassi C. Fatty acid as biological markers for bacterial symbionts in sponges 1988; 23(1) 1139-1135. http://link.springer.com/article/10.1007/BF02535280

[42] Ahmad A, Gharaibeh, and Kent J. Voorhees. Characterization of lipid fatty acids in whole-cell microorganisms using in situ supercritical fluid derivatization/extraction and Gas Chromatography/Mass Spectrometry. Analytical Methods 1996; 68(17) 2805–2810. DOI: 10.1021/ac9600767.

[43] Tunlid A, Hoitink HAJ, Low C, and White DC. Characterization of bacteria that suppress *Rhizoctonia* damping-off in bark compost media by analysis of fatty acid biomarkers. Applied and Environmental Microbiology 1989; 55(6)1368-1374. http://aem.asm.org/content/55/6/1368.short.

[44] De Carvalho, Caramujo. Lipids of prokaryotic origin at the base of marine food webs. Marine Drugs 2012; 10 2698-2714; DOI:10.3390/md10122698. www.mdpi.com/journal/marinedrugs.

[45] Fang J, Chan O, Joeckel RM, Huang Y, Wang Y, Bazylinski DA, Moorman TB, and Clement BJ. Biomarker analysis of microbial diversity in sediments of a saline groundwater seep of Salt Basin, Nebraska. Organic Geochemistry 2006; 37(8) 912–931. DOI:10.1016/j.orggeochem.2006.04.007.

[46] Lu N, Weia D, Chena F, and Yang ST. Lipidomic profiling reveals lipid regulation in the snow alga *Chlamydomonas nivalis* in response to nitrate or phosphate deprivation. European Journal of Lipid Science and Technology 2012; 114(3) 253-265. http://dx.doi.org/10.1016/j.procbio.2013.02.028.

[47] Su X, Xu J, Yan X, Zhao P, Chen J, Zhou Ch, Zhao F, and Li S. Lipidomic changes during different growth stages of *Nitzschia closterium f. minutissima*. Metabolomics 2013; 9 300–310. DOI: 10.1007/s11306-012-0445-1.

[48] Chen D, Yan X, Xu J, Su X, and Li L. Lipidomic profiling and discovery of lipid biomarkers in *Stephanodiscus* sp. under cold stress. Metabolomics 2013; 9 949-959 doi: 10.1007/s11306-013-0515-z.

Chapter 7

Applications of Modern Pyrolysis Gas Chromatography for the Study of Degradation and Aging in Complex Silicone Elastomers

James P. Lewicki and Robert S. Maxwell

Additional information is available at the end of the chapter

http://dx.doi.org/10.5772/57509

1. Introduction

1.1. Silicones — Complex and challenging synthetic polymeric materials

Silicone based materials (polymers, polymeric network elastomers, composites and hybrid materials) are a ubiquitous class of synthetic polymeric system – encountered commonly in the research, industrial and commercial areas [1]. Silicones find use in diverse areas, ranging from industrial power distribution [1], as biomedical implants [2], in the advanced electronics sector [3, 4] and the aerospace industry [5]. No one material can be considered a representative 'silicone' as the definition ranges from low viscosity liquids through compliant elastomers and gums, to intractable high modulus solids. In general however, silicones can be considered to be polymeric materials based upon an [(R_2Si)-O] repeating unit and they are most often structurally complex, multi-component systems which incorporate both chemically and physically diverse structural architectures. (See Figure 1)

It is important to understand from the outset in this chapter that gaining an understanding of the relationship between the multi-scale structure of a complex silicone system such as a commercial silicone elastomer and the resultant macroscopic properties and performance must be the goal of any informed approach to both performance and lifetime prediction of current service materials [6] and the rational development of the next generation of high performance silicone systems. [7-10]

Of all the silicone based materials in use today, it is arguably *silicone elastomers* that are both the most complex in overall structure and challenging in their analysis: Polymer structure, cure chemistry, network functionality, filler type and loading levels are just some of the

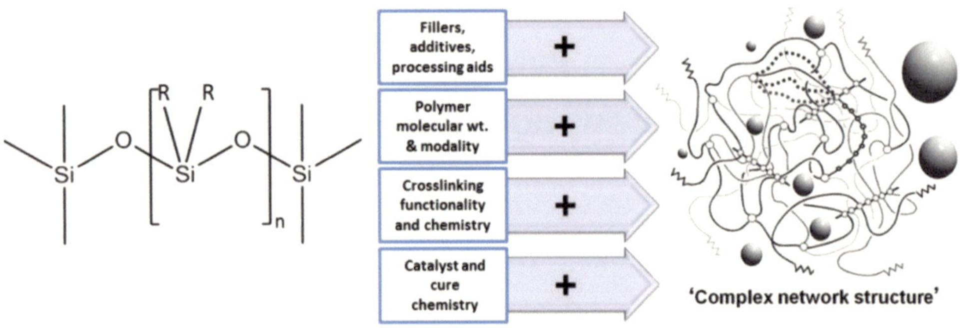

Figure 1. Left: The basic repeating structure of a polysiloxane (silicone) polymer. Right: A representation of some of the main factors that make up the overall, complex structural architecture of a silicone material.

variables that go into defining a final 3-dimensional, multi-scaled silicone 'network' elastomer. However there is a very real lack of data available on the specific structure of many important classes of silicone elastomer, due to both the proprietary nature of many industrial formulations and the simple fact that these materials are difficult to analyze at high fidelity: Silicone elastomer systems are most often chemically crosslinked networks with an effective infinite molecular weight and are as such, insoluble & *intractable*. Therefore they are inaccessible to the majority of spectroscopic and chromatographic techniques which are commonly employed for chemical Identification, characterization of structure-property relationships in polymeric systems and powerful techniques such as GCMS cannot be directly employed for the analysis of silicone elastomers. In the past these issues were not as prevalent as they are today due to the less intensive requirements of, and demands on silicone elastomers in application [11]. Simple empirically based additive methodologies were often relied upon for investigating structure-property relationships in silicone elastomers for the elucidation of chemical structure and the prediction of materials performance. [12-15] However, using these comparatively basic approaches, all but the most coarse grain information on network structure were obtained and the specifics of the 'network architecture' were largely overlooked. Today, it is accepted that the macroscopic properties and dynamic performance of a silicone elastomer system over time are subject to the multi-scale structure of the material over a range of size scales and that even comparatively small alterations to the network structure can lead to significant changes in materials properties [16, 17] (see Figure 2).

With this information in hand, it is clear that there is a need to move beyond basic 'bulk' empirical approaches to defining structure property relationships in silicones. Instead, we must employ higher resolution analytical methodologies to these materials - in order to relate the underlying network structure, (from an atomistic to a meso-scale level) directly to macroscopic materials behavior.

In spite of the intractability of silicone elastomers towards standard spectroscopic and analytical techniques, much progress has been made towards enhancing understanding of

http://dx.doi.org/10.5772/57509

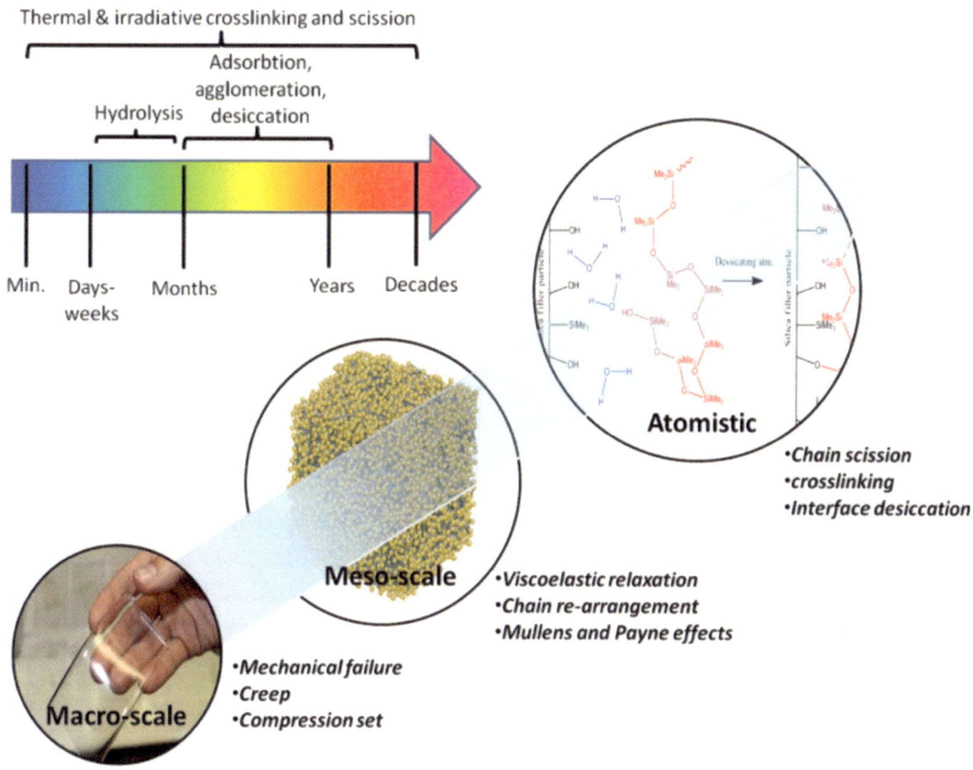

Figure 2. *Main*: Illustration of the complex relationship between materials properties/performance and structure over a range of size scale, from atomistic to a macro-scale. *Inset*: the time dependencies of various 'aging' processes that effect silicone systems in a service environment.

structure-property relations in complex siloxanes. Many researchers have made extensive use of solid state Nuclear Magnetic Resonance (NMR) both as a method of determining the chemical content/makeup of the polymer backbone in commercial silicone formulations [18] and as a tool for probing segmental dynamics of silicone networks in the solid state. [6, 19, 20] While NMR in particular has been shown extensively [21-23] as a powerful tool for elucidating the chemical identity and network architecture of complex siloxane networks, the methods employed are typically non-trivial, can be time intensive and potentially costly; requiring both significant investment in instrumentation and expertise.

Degradative thermal analysis (the thermally induced de-polymerization of a polymer or polymer network) offers an alternate route to the analysis of such complex materials. Through a destructive analysis of silicone networks and the characterization of the resultant products of thermal degradation, a surprising amount of information on the chemical identity, structural architecture and even service history of the material can be gleaned. And indeed, various incarnations of pyrolytic analysis and gravimetry have been used extensively for the analysis of 'unknown' polymeric materials for many years. [24-27] The temperature at which a polymeric material degrades, the mechanism and the products of the thermal degradation are all a function of its *underlying chemical structure* and even physical morphology. Different polymeric materials can therefore be 'fingerprinted' by their thermal or thermo-oxidative degradation behavior. Importantly for the purposes of this discussion, such degradative analyses also re-open the possibility of utilizing GC/MS as a characterization tool. However it must be noted that the complex speciation of degradation products from silicone elastomers can make the interpretation of the data with respect to the original structure challenging.

2. A silicone degradation primer

Silicones, as with all polymeric materials will chemically degrade into a range of other compounds as a consequence of exposure to thermal, radiative or chemical processes with energy sufficient to scission, oxidize, hydrolyze or otherwise break the bonds of the polymer. It is the thermal stability of a silicone system however that is most often the defining factor in its application, performance and lifetime during service. Temperature broadly affects the rates and energetic favorability of all of the major degradation processes that occur in silicone systems (with the exception of non-oxidative ionizing radiation induced scission). For example, a system that is stable to acidic hydrolysis for years at room temperature may only last weeks before failing at 75°C (see Figure 3).

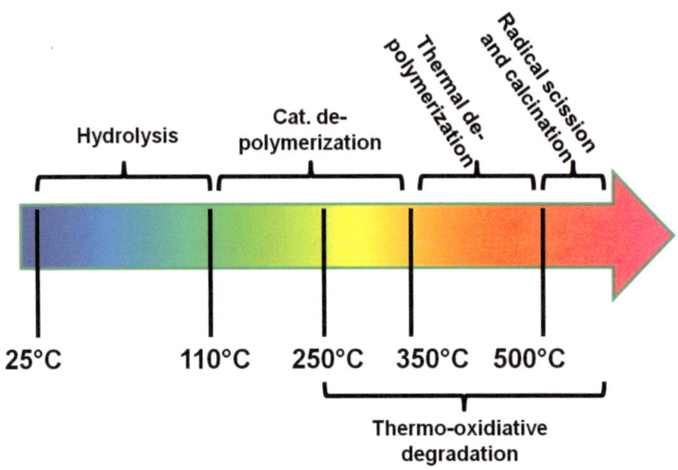

Figure 3. The primary mechanisms of silicone degradation as a function of temperature. Differing mechanisms are dominant over specific temperature ranges with more energetic process becoming dominant (within a specific time-frame) at higher temperatures.

It is unsurprising therefore, that since their introduction as usable materials some 60 years ago, there has been interest in the thermal stability and the mechanisms by which silicones degrade at elevated temperatures. The most influential work on the subject of the thermal degradation of the silicone polymers was carried out by Grassie et al. [28-35] who studied in detail the thermal and thermo-oxidative degradation of a range of linear silicone polymers using a combination of Thermal Volatilization Analysis (TVA) and Thermogravimetric Analysis (TGA). Building upon earlier work by Pantode and Wilcock [36] and Thomas and Kendrick, [37, 38] Grassie demonstrated through an in-depth analysis of the products of thermal degradation that silicones, in general, at elevated temperatures degrade via a de-polymerization reaction to yield cyclic oligomeric siloxanes. This thermal de-polymerization reaction proceeds from both free chain ends and as a result of intra-molecular backbiting reactions of continuous chain segments (see Figure 4). For linear poly(dimethylsiloxane) (PDMS) Grassie and Macfarlane reported the major degradation products to be cyclic oligomeric siloxanes of ring sizes D_3–D_{12} and higher oligomeric siloxane species.

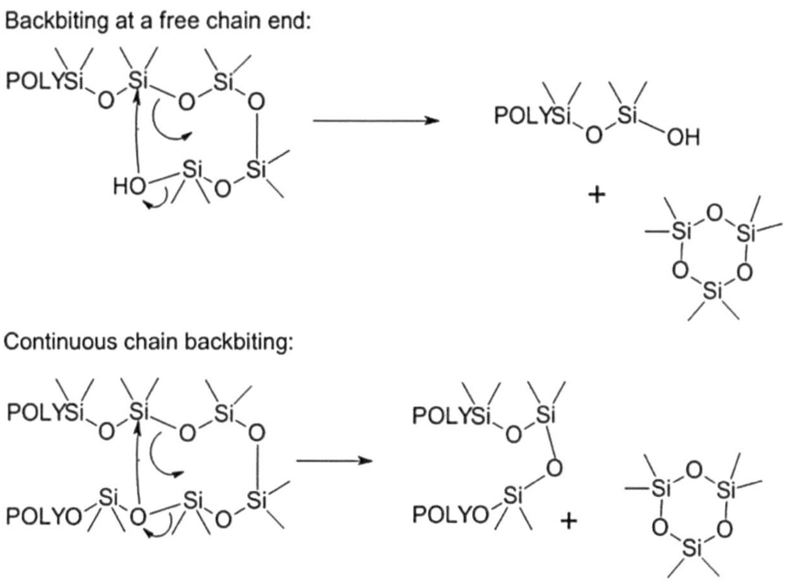

Figure 4. Intramolecular chain backbiting mechanisms. Upper scheme illustrates the attack of a chain by its own OH terminated free end to form a cyclic siloxane. Lower scheme illustrates the case were a closed, continuous chain folds back upon itself and re-arrangement occurs to form a new Si-O bond and a free cyclic siloxane

Uncatalyzed backbiting and cyclization reactions such as those proposed by Grassie in Figure 4 are typically reported to occur at temperatures of 350-400 °C and are in essence, the reverse of the ring-opening polymerization reactions that are employed to synthesize high molar mass silicones from cyclic oligomeric silicone pre-cursors. From a thermodynamic standpoint the ceiling temperature for the polymerization of cyclic silicones to form PDMS is relatively low (~110 °C). The implication is therefore that the reverse reaction – de-polymerization to reform cyclic oligomers is thermodynamically favored above temperatures of 110°C. However as Grassie demonstrated [28] in the late 1970s even linear unmodified PDMS in the absence of catalyst residues or impurities is stable up to a temperature of 350°C and in practice, a silicone can rarely be expected to de-polymerize at a significant rate (unaided) at temperatures below 250 °C. It is therefore reasonable to assume that the influence of kinetic, steric and other factors in real materials all contribute to higher experimentally observed thermal stabilities of silicones.

Backbiting cyclization reactions are recognized in the literature [28, 39] as one of the primary thermal degradation mechanisms that occur in silicone systems. Importantly, however these backbiting reactions become significantly more favorable in the presence of acid or base. [28, 40] The presence of catalytic levels of Lewis acids or bases is now known to accelerate the depolymerization reaction and lower the degradation temperature of silicones significantly. Such catalyzed backbiting reactions are dominant over a temperature range of 110-260 °C. Silicones are also highly susceptible to hydrolysis: The ([R_2Si]-O) bond is comparatively strong, having an average bond disassociation energy of 452 KJ mol^{-1}, it is strongly polar and sterically unhindered, making the Si center highly susceptible to nucleophillic attack. A practical consequence of this nucleophillic susceptibility is that silicones are readily hydrolyzed under mildly acidic or basic conditions. At temperatures below 110 °C, thermal hydrolysis is one of the most significant processes that contribute to the long term aging and failure of silicone elastomers. [41]

Today, the thermal degradation chemistry of simple, linear siloxane systems is relatively well understood. However, many commercial siloxane polymer systems are significantly more complex. Such systems may contain inorganic fillers, crosslinking agents, catalysts, processing aids, synthesis residues, curing reaction by-products and other chemical 'complications' – all of which can alter their degradation behavior significantly. [42-46] Such systems are non-trivial to both analyze structurally, predict the properties of long-term and are the subject of much research today. [5, 27, 47-49]

3. Analytical degradative thermal analysis

3.1. Introduction

There are a number of degradative thermal analysis techniques available for the analysis of silicones which include Thermal-Gravimetric Analysis (TGA), Differential Scanning Calorimetry (DSC) Pyrolysis-Gas Chromatography/Mass Spectrometry (Py-GC/MS). These methods (and others), their application and the broader field of thermal analysis of polymers have been reviewed thoroughly in the excellent text by Wunderlich. [50] However, for the purposes of discussing analytical thermal methods for the determination of structural information from complex silicone networks, we will focus ourselves only on those techniques capable of providing chemical information on the products of silicone degradation as a function of temperature with the potential for quantification. Specifically, pyrolysis methods coupled with chromatography/spectrometry/scopy - such as Py-GC/MS.

Silicones degrade to form a range of degradation products and it has been established that both gross alterations in the chemical nature of the silicone polymer [28-35] and relatively subtle alterations in the network architecture of a silicone material [51, 52] can alter the speciation of products formed. It follows therefore that the purposeful, analytical degradation

http://dx.doi.org/10.5772/57509

of a silicone material coupled with the subsequent analysis of the products of degradation may allow the elucidation of an unknown network structure and even enable forensic identification (fingerprinting) of unknown engineering elastomers by their degradation profiles. [53]

3.2. Vacuum pyrolysis-mass spectrometry

Analytical pyrolysis (the controlled thermal degradation of organic or inorganic materials with the subsequent identification of the gas-phase products of degradation) has long been employed for the study of polymeric materials degradation. [26] In general, analytical pyrolytic methodologies involve the rapid thermal degradation (>100 °C/min) of a polymeric sample to complete thermal degradation under a controlled atmosphere. The effluent gas stream is analyzed by a secondary chromatographic and/or spectroscopic technique to identify the products of the thermal decomposition of the material. Pyrolytic methodologies rely on small sample masses, high analyte detection sensitivity and operate in a non-diffusion limited regime (See Figure 5).

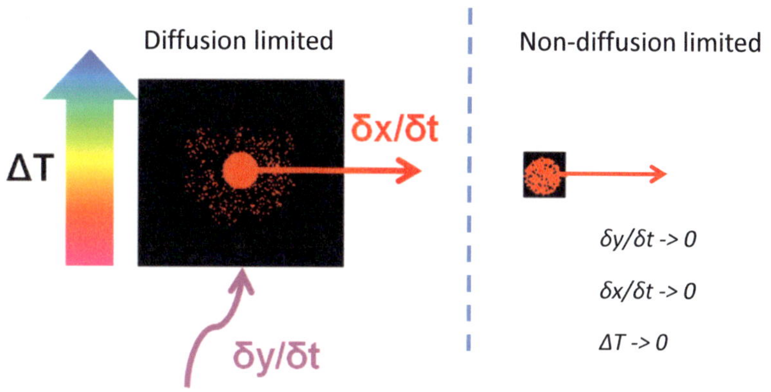

Figure 5. In bulk degradation analyses (5-50 mg sample size) (left) the sample mass requirements and the limitations of experimental geometry give rise to significant diffusion and thermal gradients, rates of diffusion for reactants into the materials and products out (*dy/dt* and *dx/dt*) are often ignored despite their importance in any kinetic or mass transport model derived from the data. A technique that allows the use sub-milligram samples or thin films, sidesteps many of these complications - simply on the basis of surface to volume ratio (right).

Pyrolytic analysis methodologies are therefore typically regarded as more versatile, less biased, chemical information rich techniques which can rapidly and more reliably access the mechanistic degradation behavior of polymeric materials. Early applications of what would later become analytical pyrolysis of silicone systems include work by Wacholtz et. al [54] in which the pyrolysis products of the thermal degradation of a silicone system were sampled and analyzed in-line using a combination of FTIR and GC/MS. Such early in-line pyrolysis studies of silicone degradation would later form the basis of modern micro-analytical pyrolysis methodologies for the analysis of silicone degradation. However, until comparatively recent times the field of analytical degradative analysis of silicones was almost entirely dominated by the vacuum pyrolysis technique –Thermal Volatilization Analysis (TVA).

TVA is essentially an evolved gas analysis technique which is based upon the principle of accurate measurement of the pressure of volatile species evolved from a material undergoing a heating regime. TVA effectively monitors the evolution of volatile degradation products of a sample as a function of pressure vs. temperature as the sample is subjected to a linear heating ramp under vacuum. The technique was originally developed in the 1960s by Ian McNeill and co-workers [55] as a tool for studying polymer degradation. Being related to other vacuum based thermal analysis techniques used at this period to study polymer degradation, [56, 57] it rapidly became popular amongst polymer degradation groups and was utilized in many of the seminal studies of polymer degradation. [28] There were several incarnations of the basic technique and it saw its most advanced development in sub-ambient thermal volatilization analysis (SATVA) - which combined the differential monitoring of degradation behavior as a function of pressure, with the collection and separation of condensable volatile products by cryo-trapping and sub-ambient differential distillation. A general schematic diagram of a TVA system is shown in Figure 6.

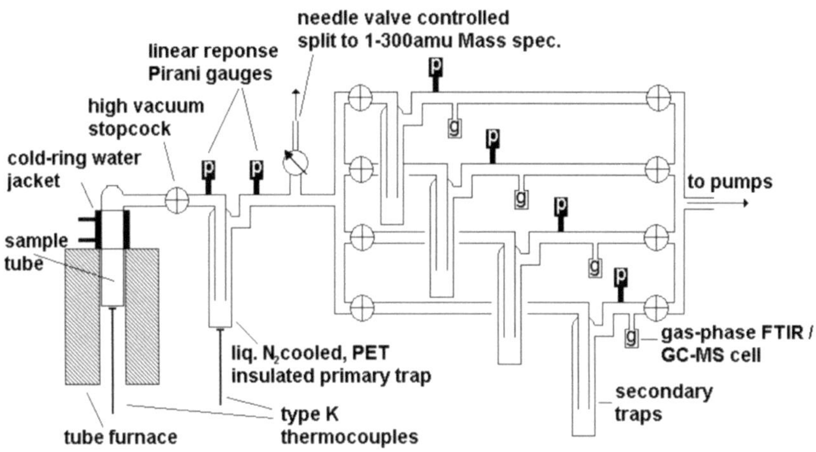

Figure 6. A general schematic representation of a TVA vacuum pyrolysis system

The apparatus consists of a sample chamber (heated by a programmable tube furnace) connected in series to a primary liquid nitrogen cooled cryo-trap and a set of four secondary cold traps. The system is continuously pumped to a vacuum of ~1x10^{-4} torr. Volatile condensable products can be initially trapped at two stages: The water jacket cooled 'cold-ring' (T ~ 12 °C) immediately above the heated area of sample tube which condenses high boiling point materials and the main cryo trap (T ~-196 °C) which collects all the lower boiling point condensable species. Two linear response pressure gauges are positioned at the entrance and exit of the main cryo-trap to monitor the evolution of both condensable and non-condensable volatiles as a function of pressure vs. temperature/time from the sample. The primary cryo-trap captures all of the condensable lower boiling point species evolved during the pyrolysis of a sample. This trapped condensate can then be distilled into separate secondary liquid nitrogen cooled cold traps by controlled heating of the primary trap to ambient or elevated temperatures. A mass spectrometer samples the gas stream continuously at the exit of the primary cryo-trap, providing a means of identifying non-condensable species such as methane or hydrogen evolved from a sample in addition to mass spectral identification of products in as they are separated during the differential distillation stage. Data from the mass spectrometer can be correlated with the sub-ambient distillation pressure peak plots and greatly aids in the identification of volatile species. Distilled product fractions can be subsequently removed into gas-phase cells for offline FTIR and GC/MS analysis. A series of secondary pressure gauges are placed at the entrance and exits of all secondary traps to monitor the distillation of specific product fractions into separate traps and gas cells.

The TVA technique is a vacuum pyrolysis method, coupled with a thermal trap/desorption system for volatile product separation and in many ways it is a predecessor to the modern pyrolysis-gas chromatography systems commonly in use today. The relative complexity both in design and operation of the method, along with the molecular weight range and resolution limitations of vacuum distillation separation prevented TVA and related vacuum pyrolysis 'line' systems from ever becoming more than a niche technique. And with the introduction of commercial gas chromatography systems directly coupled to fast quadrupolar mass filters throughout the 1970-1980s – the technique has been largely superseded by simpler and more versatile in-line pyrolysis GC/MS systems. That being said, a limited number of academics still utilize TVA as a means of studying polymer degradation [58-60] and indeed, it has even been used in recent years for the study of complex silicone nanocomposite elastomer systems. [51, 61] One such example is given in Figure 7.

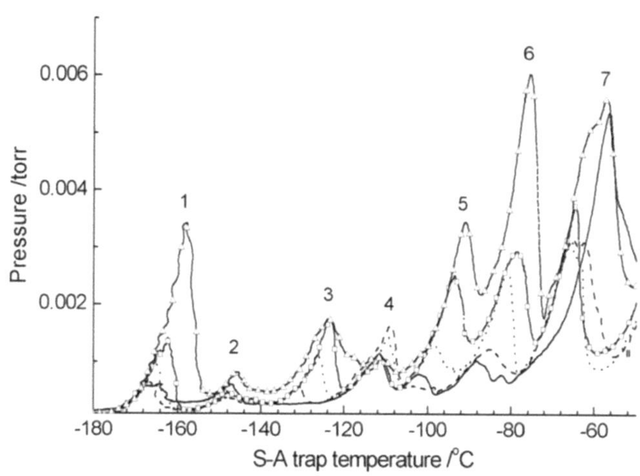

Figure 7. Sub-ambient differential distillation traces of the captured volatiles from the thermal degradation of an un-filled silicone and a range silicone-nanoclay composite systems. Solid, dashed, dotted, squares & circles represent 0, 2, 4, 6, & 8% clay filled composites respectively. Degradation products elute as a function of their boiling point as the trap is heated at a rate of 10 °C/min. The peaks labeled 1-7 were identified by a combination of gas-phase MS, gas-phase FTIR and GC-MS and are listed in Table 1. Reprinted with permission from. [51] Copyright Elsevier (2009).

Peak number	Product ID	Relative Abundance	Detector	Source
N/A non-condensable	Methane	minor	MS	high temp radical reactions
1	Propene & CO_2	trace	MS/FTIR	elimination reaction of x-linker residues
2	butene	minor	MS/FTIR	catalyst residue elimination
3	the enolate form of dimethylsilanone	trace	MS/FTIR	main chain radical scission
4	Linear silicone –	minor	MS	x-linker residues
5	Propanal	trace	MS/FTIR	scission of chain end modifier
6	Benzene	minor	MS/FTIR	catalyst residue elimination
7	Ethyl hexanoic acid	minor	FTIR/GCMS	Catalyst residue (intact)
8 (not shown due to scale)	D_3-D_5 cyclic oligomers	major	GCMS	thermal backbiting reactions
N/A semi-volatile	Higher oligomeric siloxanes	major	GCMS	thermal backbiting reactions & equilibration

Table 1. Identified products from the TVA analysis of a range of silicone nano-composite elastomers.

From the TVA data given in Figure 7 and Table 1 it is clear that while the quality of the chromatography from differential distillation is poor compared with GC/MS, there is still much data that can be obtained from such analyses. The speciation and distribution of products observed in this particular study allowed the authors to determine that a clay filler additive in a particular silicone network significantly changes the speciation of minor products formed via catalyzing secondary racial scission reactions at high temperatures.

3.3. Modern micro-pyrolysis-GC/MS

A more versatile approach to the analytical degradative analysis of the complex silicone based materials is modern Micro-Pyrolysis-coupled Gas Chromatography / Mass Spectrometry. More commonly referred to as Pyrolysis GC/MS, this technique was developed by and grew popular in the materials research and the oil industries. Py-GC/MS paralleled the development of commercial GC systems and found extensive application in the study of materials degradation and destructive analysis of intractable organic materials. The technique today remains disarmingly simple: a low thermal mass oven which is directly in-line with a detector (GC/MS etc.) heats a very small sample of analyte rapidly under inert or reactive carrier gas to destruction and the volatile products of degradation are passed by means of the carrier gas for analysis and optional separation by the detection instrument (see Figure 8).

Owing to the fact that the effluent stream of the furnace can be directly fed into the inlet of a sensitive GC/MS system, modern pyrolysis requires only extremely small analyte masses; from $1x10^{-3}$ to 1 mg is typical for an analytical pyrolysis of an organic containing material. Added to that, the thermal mass of a pyrolysis furnace is typically low and often in the form of a platinum coil/quartz tube 'probe' type geometry (See Figure 9). A broad range of heating rates

http://dx.doi.org/10.5772/57509

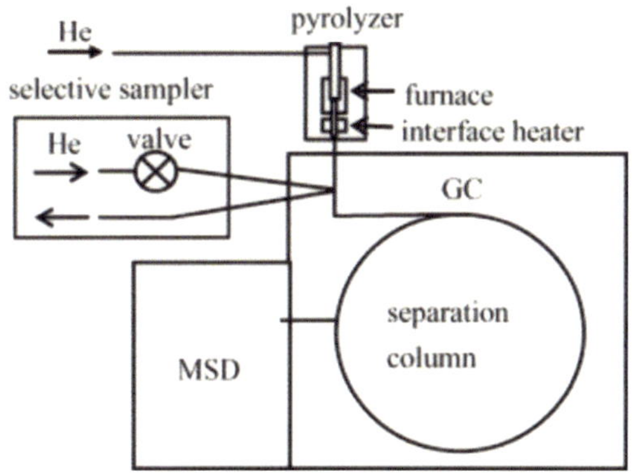

Figure 8. Schematic representation of an inline pyrolysis GC/MS system. Reprinted with permission from. [62] Copyright Elsevier (2006).

and a high maximum temperature are therefore accessible (0.01 to 999°C/min with a maximum of 1200 °C are typical).

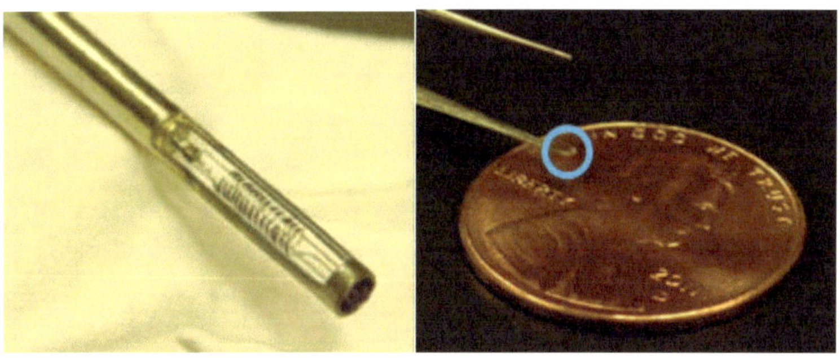

Figure 9. On the left is a modern pyrolysis probe showing the platinum coil heater and quartz sample tube. Right is an example of the typical mass of a silicone elastomer (highlighted by the blue circle) required for an analytical degradative analysis.

The ability to analyze such small samples of material; importantly allows degradation and offgassing processes to be studied in a more ideal non-diffusion limited regime, negating the limitations of bulk analysis techniques such as TGA. One of the other great advantages of analytical Py-GC/MS is that it can provide both hi-fidelity, in-depth chemical speciation data in addition to rate/loss thermogram type data simply by means of changing the elution conditions of the GC column (see Figure 10).

Using pyrolysis-GC/MS we can monitor the degradation of a silicone system in real time (as in the left plot of Figure 10) to yield assessments of thermal stability, overall degradation profile and level of volatiles released. For example, in this study it was observed that the level of volatiles released from a silicone elastomer decrease markedly with increasing carbon nanotube content. Importantly, however, we can also obtain full product separation and identifi-

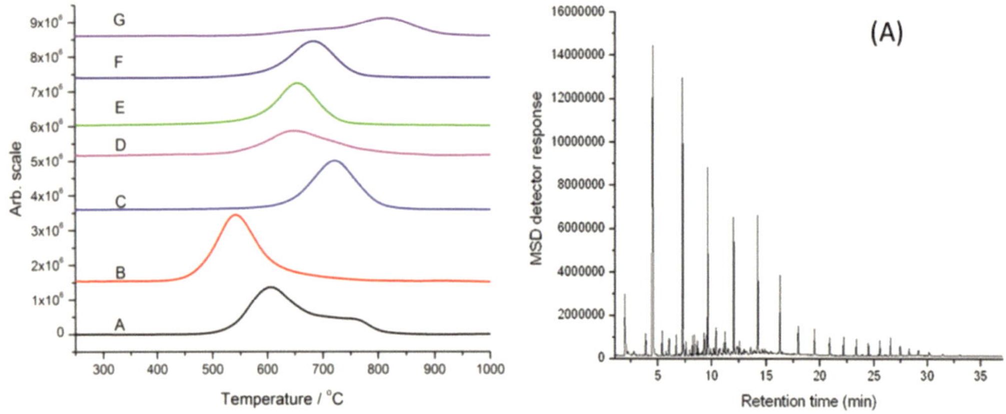

Figure 10. Degradative analysis of a series of silicone-carbon nanotube composites using Py-GC/MS. Left: pyrolysis thermogram data obtained by pyrolysing a series of silicone composites at a ramp rate of 100°C/min while using the GC/MS in a non-separating direct detection mode. A-G corresponds to silicone formations with increasing levels of carbon nanotubes. Right: full product speciation obtained from the ballistic pyrolysis of sample 'A' at 999 °C/min with the subsequent separation and speciation of the degradation products by GC/MS.

cation for a given degradation process using the same apparatus, now operating in the more conventional flash or 'ballistic' heating mode with full GC/MS separation and detection of analytes (right hand Figure 10). In this particular study it was shown from an analysis of the degradation products provided by Py-GC/MS that while the carbon nanotube additive did not significantly change the chemistry of the silicone degradation, it did physically retard the release of volatile species and increase the yield of calcined 'char' – thus leading to improved thermal stability.

3.4. Quantitative Py-GC/MS methods

As with the majority of GC based techniques, Py-GC/MS can through calibration, be fully quantitative with respect to the outgassing or degradation species being monitored. Through the use of an external calibrant standard (in the following example, analytical grade $NaHCO_3$) the instrument response to various gaseous species as a function of the total number of moles evolved from the pyrolysis event can be used to build a calibration curve. See Figure 11.

In the relatively simple example shown in Figure 11 varying masses of $NaHCO_3$ were pyrolysed under He in-line to 400 °C. This rapidly decomposed the bicarbonate to carbonate, CO_2 and H_2O in a fixed stoichiometric ratio. The pyrolysates were trapped on a cryo-cooled column to collimate them and then allowed to elute through the detector separately. Both ions 44 and 18 were monitored during chromatographic separation to yield a detector response as a function of moles of CO_2 and H_2O. With enough data points over a suitable mass range, instrument/method dependent calibration curves can be constructed which allow quantification of analyte gasses from unknown samples.

http://dx.doi.org/10.5772/57509

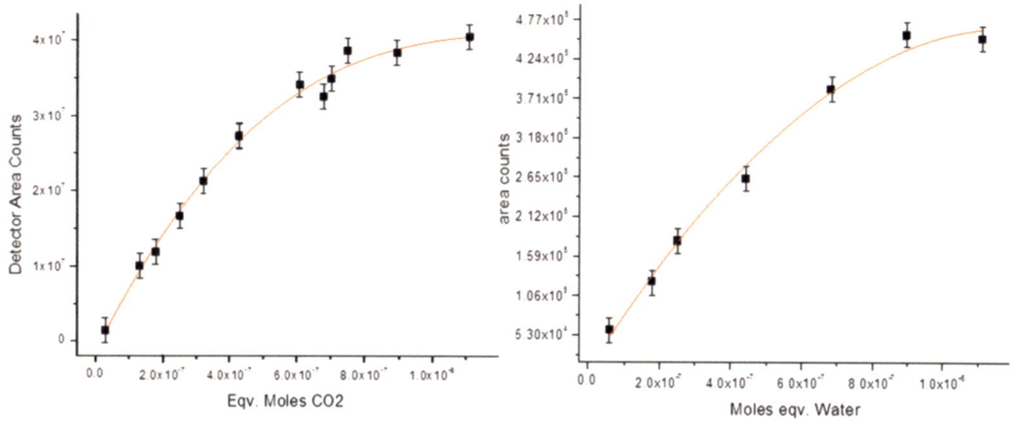

Figure 11. Py-GC/MS calibration curves obtained from the pyrolysis of known quantities of $NaHCO_3$. Both ions 44 and 18 were monitored during chromatographic separation to yield a detector response as a function of moles of CO_2 and H_2O.

Indeed, these data have been used in the following example to quantify the diffusion of CO_2 through a silicone membrane directly and as a function of temperature using Py-GC/MS. Shown in Figure 12 are the results of quantitative Py-GC/MS analysis of the diffusion of a sealed, finite CO_2 source through a 250 μm thick silicone membrane at a temperature of 120°C as a function of time.

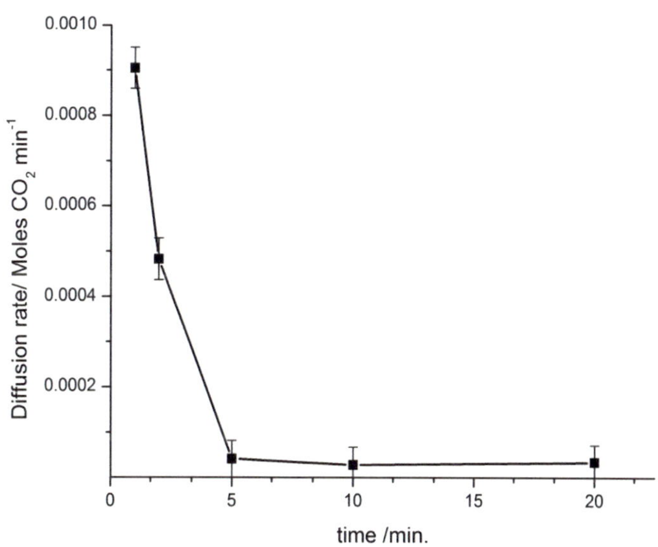

Figure 12. Evolution of CO_2 from a finite source, as measured by Py-GC/MS through a thin silicone membrane at elevated temperatures (quantitative on CO_2). Note that the initial rate of diffusion through the membrane is rapid but as the CO_2 source is depleted, a sharp decrease rate is observed.

4. The application of chemometric techniques to large Py-GC/MS datasets

Pyrolysis GC/MS data from silicone systems can be complex, with ion chromatograms from a single degradation or off-gassing process often having many hundred eluted products and a single study may yield many such individual chromatograms which must be interpreted. Often, therefore we must rely on more advanced data processing and mining techniques to both analyze these data in a timely manner and ensure that significant trends or processes are not lost in the sheer volume of data produced. A comparatively simple, yet highly effective chemometric technique that lends itself well to the analysis of such large chromatographic datasets is multivariate statistical analysis – specifically, Principle Components Analysis (PCA).

In PCA as with many other multivariate statistical techniques, sample-to-sample or object-to-object variation can be represented by a series of principal components (PCs), which preserve the structure of the underlying variance between two or more variables. The general aim of PCA is the reduction of the dimensionality of a data set by the computation of a small number of these components (typically much less than the number of variables) that are parameterized by so-called scores and loadings. See Figure 13.

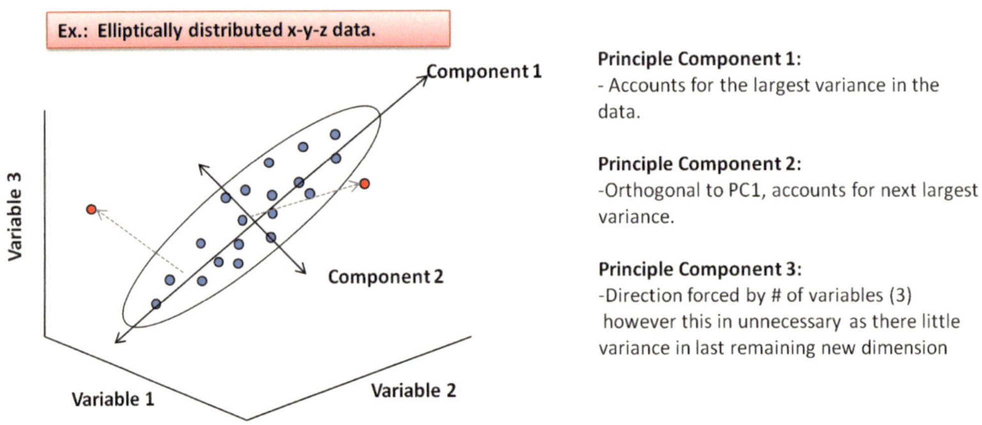

Figure 13. An illustration of the dimensional reduction that allows PCA to de-convolute large datasets. In this example there is elliptically distributed x, y, & z data (blue). Through an orthogonal coordinate rotation the largest variances in the data can be separated into principle components each with their own dimensionality. At the same time we reduce the overall dimensionality of the dataset by separating and discarding unwanted variables such as spectral noise. Finally, outliers are clearly identified outside of confidence limits in each PC.

Each component derived from PCA contains within it some proportion of the overall variance (generally expressed as a percentage of the total), with the first principal component (PC1) being the latent variable which describes the maximum amount of variance over a given dataset. PC2, the second principal component, is uncorrelated with (orthogonal to) the first PC and accounts for the second largest percentage of the overall variance. Additional PCs can be defined similarly. For the pyrolysis data considered presently, the loading vectors profile the extent of retention time, with peaks representing regions of significant variance among samples. Scores provide information about the degree to which a given loading is important for a particular sample/chromatogram. In other terms, scores can be thought of as weights or as "concentrations" of loadings for the latent variables. [63]

PCA analysis has recently been applied to a number of studies of silicone degradation. [5, 52, 53] And in the first of two such studies we will discuss, the thermal degradation behavior of a series of well-defined model poly (dimethylsiloxane) model networks were investigated with Py-GC/MS and PCA in order to probe the influence of controlled network architectures on degradation chemistries and product profiles. A matrix of model silicone networks were formulated to incorporate a range of well-defined network architectures. Specifically; monomodal over a range of chain molar masses above and below the critical entanglement molar mass, M_c (~12 KDa); bimodal with varying mol. percentages of short (8 KDa) and long (133 KDa) chains; free chain end containing (a monomodal network with 1-20 mol. percent free chain ends). Samples were pyrolysed at a ballistic heating rate to 1000 °C under helium. The products of degradation were analyzed using in-line GC/MS to yield total ion chromatograms. Shown in Figure 14 is an example of a typical GC trace obtained from the pyrolysis of one of the model silicone networks.

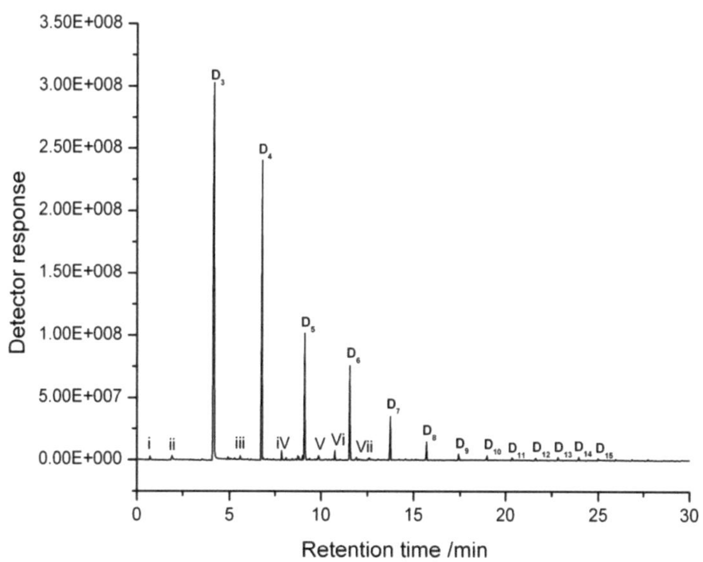

Figure 14. GC total ion chromatogram (TIC) of the pyrolysis products from a 0.04 mg sample of a 54.4 KDa monomodal crosslinked PDMS elastomer. The products of pyrolytic degradation are labeled D_3 to D_{15} (cyclic siloxanes) and i-vii (misc small molecule, branched and linear species). Due the large size of the datasets obtained from such a study (>100 high resolution chromatorgrams each with 20-40 identified products) PCA was applied to the datasets. Shown in Figure 15 is a 'samples and 'scores' plot for principal components 1 & 2 of a PCA model capturing 90% of the total variance in the complete matrix of silicones studied.

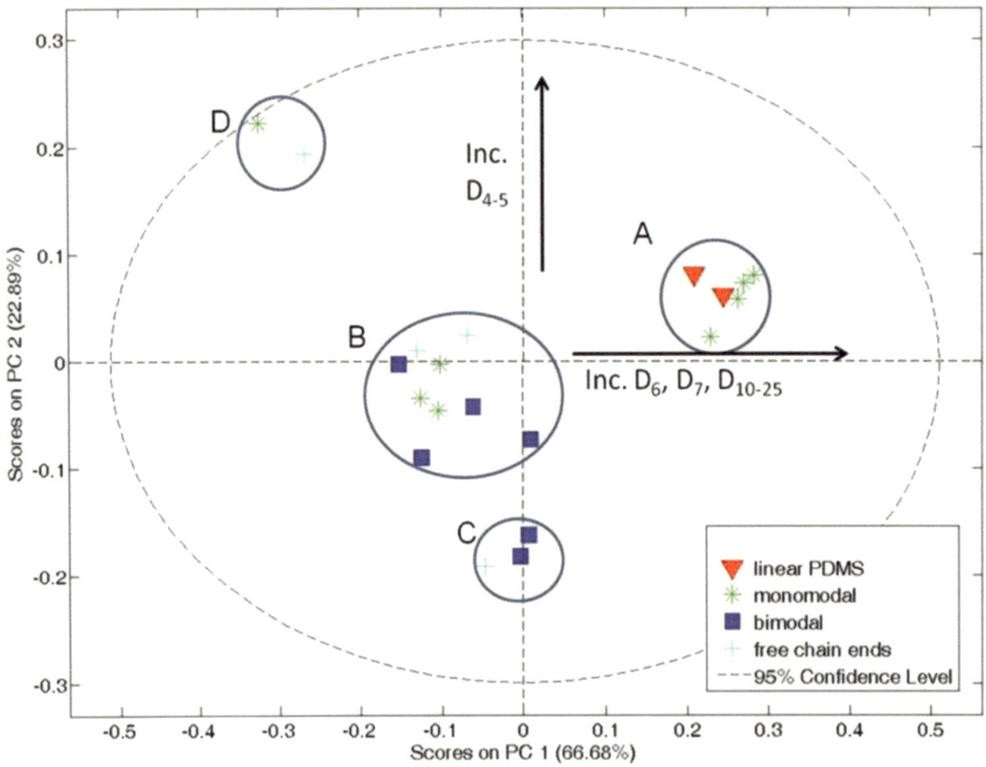

Figure 15. Samples-scores plot for PC's 1&2, complete sample matrix. The matrix had been grouped into four classes of data, corresponding to the linear PDMS, the monomodal, bimodal and free chain end subsets respectively. Groupings of samples are circled and labeled A-D.

From Figure 15 it is apparent that there are a number of significant groupings of samples (labeled A-D) suggesting that samples within these groupings degrade in a similar manner. Positive scores on PC1 (x-axis) have been correlated with increased relative yields of D_6, D_7 & $D_{10\text{-}25}$ cyclics. Positive scores on PC2 have been correlated with increased relative yields of D_4 & D_5 cyclics. These correlations are obtained from the analysis of variable 'loadings' plots as shown in Figure 16.

Loadings data such as the example shown in Figure 16 allow the variables responsible for a component to be readily examined. Here positive loadings correlate with increases in D_4 & D_5 cyclics and negative loadings relate to decreases in methane, propene levels and D_6 cyclics

From an examination of the scores and loadings of a valid PCA model we can make confident assertions as to both global trends and specific changes in degradation chemistry. From an examination of our PCA model in this example it can be observed that global degradation behavior of a large series of model silicones can be readily mapped using PCA. Trends made apparent through scores analysis can be related back to the underlying chemical information by means of the variable loadings data. These relations can in turn be used to correlate a specific chemical degradation behavior to the underlying network architecture. For example, Group

http://dx.doi.org/10.5772/57509

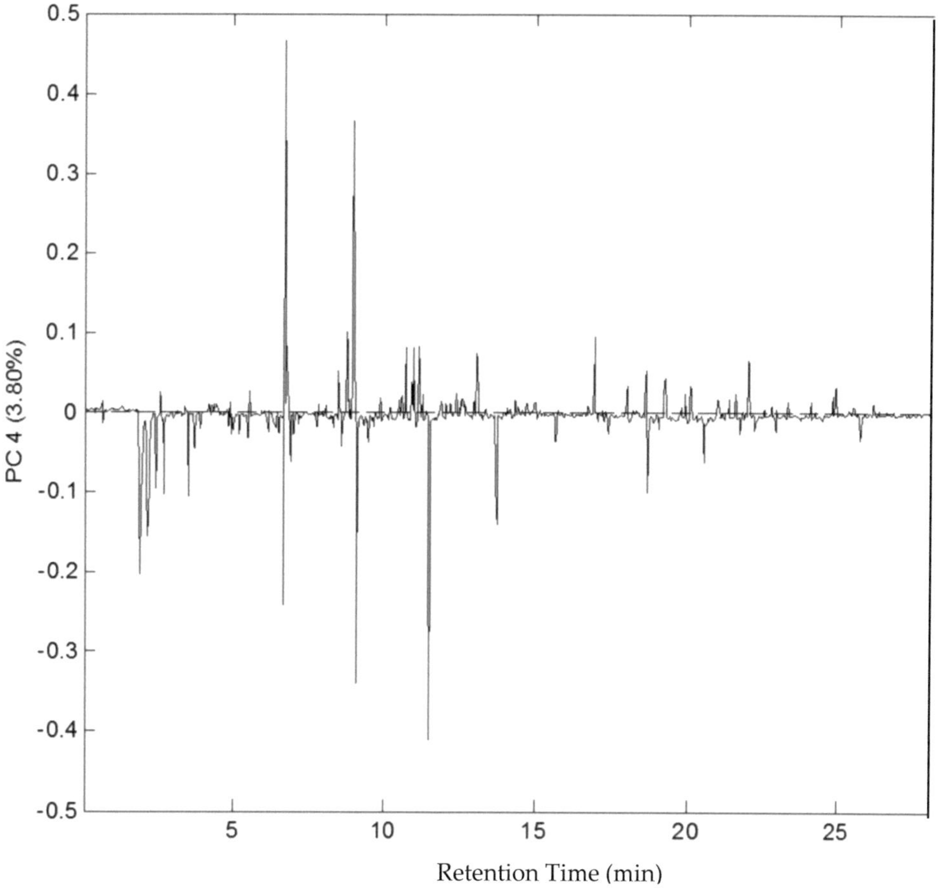

Figure 16. A loadings plot generated from a PCA model of silicone degradation pyrolysis data. Shown for example here is are the variables responsible for the 4th principle component of the model.

'A' in Figure 15 includes both 8 and 132 KDa linear PDMS and a series of 8, 10, 33 and 54 KDa monomodal networks. The strong positive score on PC1 for this group indicates that they yield increased levels of larger cyclic siloxanes relative the rest of the matrix. The samples in group 'B' include 1 and 10% free chain end samples, a 68 KDa monomodal material and a series 90, 80, 70 & 50% short chain bimodal networks. Group 'B' is representative of the mean degradation behavior of the model networks. Group 'C' includes a 5% free chain end, 80% and 95% short chain bimodal systems. Group 'C' displays broadly similar behavior to the mean group (B) however there are a somewhat reduced level of $D_{4\text{-}5}$ cyclics produced in this group of samples. The final group 'D' includes the 132 KDa monomodal sample and the 20% free chain end sample. Both of these samples evolve significantly reduced quantities of larger cyclics and increased levels of smaller $D_{4\text{-}5}$ cyclic siloxanes.

In the first example we demonstrated that Py-GC/MS methodologies can effectively discriminate between specific network architectures in simple model silicone systems as a function of their degradation chemistry. Significantly, it has also recently been demonstrated [53] that valid structure property correlations can be drawn from the pyrolytic analysis of complex, real-world silicone elastomers. In this second example the thermal degradation behavior of a series of commercial and specialty silicones are compared, using a combination of py-GC/MS and PCA. A summary of the formulations analyzed are given in Table 2.

Elastomer	Base polymer(s)	Network Modality	Filler type and loading level	Cure Chemistry
DC 745 (Dow Corning)	Polydimethylsiloxane Polydiphenylsiloxane	Bimodal	30 Wt. % SiO_2 (Mixture of high surface area fumed silica and low surface area quartz)	Free radical vinyl addition cure Organic peroxide catalyst.
TR-55 (Dow Corning)	PDMS Polydimethylhydrosilane Polymethylvinylsiloxane	Tri-modal	15-40% trimethylated silica	Free radical vinyl addition cure Organic peroxide catalyst.
M97 (LLNL)	PDMS PDPS PMVS	Bimodal	21.6% Cab-o-Sil M7-D silica 4% Hi Sil 233 silica	Free radical vinyl addition cure Organic peroxide catalyst.
Sylgard ® 184 (Dow Corning)	PDMS PMHS	Bimodal	30% Vinyl functionalized Silica	Platinum mediated vinyl addition cure Platinum (0) complex catalyst
S5370 (Dow Corning)	PDMS PMHS	Bimodal	30% diatomaceous earth silicate filler (Celite 30 B)	Silanol-silane condensation cure Stannous Octanoate Catalyst

Table 2. Formulation reference data for each engineering silicone elastomer system studied categorized by base polymer, final network modality, filler type/loading and cure chemistry. Reprinted with permission from. [53] Copyright Elsevier (2013).

Using standard micro-pyrolysis methodologies, these silicone elastomers were analyzed and with the application of PCA, the following data were obtained; shown in Figure 17 is the global PC1&2 scores plot encapsulating >95% of the total variance in the complete dataset under a single four component model.

PCA of the complete pyrolysis dataset show distinct groupings (labeled A-D in Figure 17) of various elastomer types. These groupings can be correlated with aspects of their underlying network chemistries: The S5370 samples (A) are clear outliers due to the fact that they are the

http://dx.doi.org/10.5772/57509

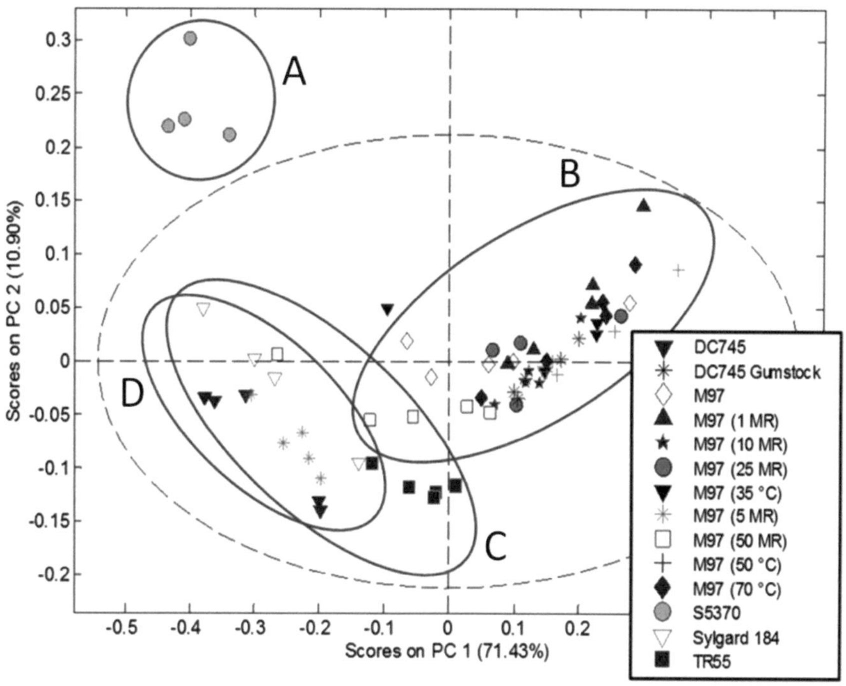

Figure 17. Global PC1 vs. PC2 Scores plots from the PCA of pyrolysis data of a series of commercial silicone elastomer formulations. The center of the plot at 0,0 can be considered represent the 'mean' of the degradation profiles averaged across the dataset as a whole. Solid lines are drawn for emphasis of groupings and labeled A-D and the dotted line represents the 95% confidence interval. Reprinted with permission from. [53] Copyright Elsevier (2013).

only system that is condensation cured and therefore significantly different in formulation to the other systems studied. In grouping (B) the M97 systems are observed to broadly group together on one axis *irrespective of their thermal or irradiative history*. This group of elastomers is in whole positive in PC1, however scattering is observed in the more extreme aged/irradiated samples; suggesting that the starting network structure is a more significant factor in determining the degradation behavior than any subsequent environmental factor. Moreover, the M97 series despite being peroxy cured addition networks and thus similar in chemistry to the majority of the other systems, remain distinct from the related commercial materials (TR55, DC745 and Sylgard 184). This distinction demonstrates that chemically similar silicone elastomers can be differentiated by means of pyrolysis GC/MS and multivariate analysis. Grouping (C) encompasses the behavior of TR55, which is neutral with respect to PC1 and negative on PC2. Finally grouping (D) highlights a correlation between The DC745 and Sylgard 184 degradation profiles.

It can therefore be shown that it is possible to link these distinct degradation 'fingerprints' to underlying chemical features of the networks themselves. The elastomers showing negative PC1 scores include those systems that are peroxy-cured: DC 745 and TR-55. DC 745 and an uncured DC 745 gumstock also group significantly in the lower left quadrant and are correlated with Sylgard 184, a Pt catalyzed addition cured system (C-D). The most notable grouping on the PC1 vs. PC2 scores is that of S5370 (group A). Notably, S5370 is the only system which employs tin catalyzed condensation crosslinking chemistry [27] rather than radical or Pt mediated addition crosslinking. As such, the network structure of S5370 has no alkyl linkages at crosslinks and is therefore significantly different to that of all the other materials. S5370 is also only one of two materials in this study that are foams (the other being M97). There is however no correlation between M97 and S5370 on the PCA map, suggesting that the physical foam structure of these materials is not the predominant factor in determining their degradation behavior. The distinct S5370 grouping therefore appears to reflect a significantly different degradation behavior as a direct result of its differing network chemistry (a fully [Si-O] based network which retains an active tin catalyst residue, capable of promoting degradation reactions). [46, 64, 65]

The data clearly show that the degradation product profiles of the studied elastomers fall into distinct observable groupings which appear to correlate directly to major features of these systems formulation chemistry, e.g. the network chemistry and catalyst type. The physical structure of the materials (fully dense vs. cellular) does not appear to be a major factor in the degradation behavior; an observation that is consistent with the fact that microgram scale ballistic pyrolytic degradation occurs in a non-diffusion limited regime and should remain broadly independent of bulk sample effects. All of these observations are both notable and consistent with what is known about silicone network thermal degradation: Addition cured networks contain alkyl linkages and are as such not purely [Si-O] networks. Both the peroxy and Pt cured systems retain no active catalyst residues (unlike the tin cured systems) therefore backbiting thermolysis and catalytic de-polymerization reactions may be expected to occur at differing rates and with differing favorability to those of a purely [Si-O] based network which retains an active catalyst residue.

5. Conclusion

The aim of this chapter has been to demonstrate that analytical degradative analysis, when supported by modern gas-chromatography techniques is an effective and versatile tool for the in-depth study of highly complex, yet important silicone elastomers systems that are challenging to analyze with other established techniques. The case studies and reviews discussed in this chapter, demonstrate that silicone materials formulated via differing cure chemistries have distinct degradation fingerprints observable by means of analytical pyrolysis. The application of PCA statistical methodologies to Py-GC/MS data allows these unique signatures to be rapidly and reliably identified. Furthermore, PCA allows the chemical origins of the degradation fingerprints to be assessed with comparative ease. The structural architecture of a network elastomer and crosslinking chemistries employed in its formation can be related to its degradative behavior. It has also been demonstrated that the analytical pyrolysis methodologies currently employed are *insensitive* to those comparatively subtle chemical and physical alterations to a network as a result of thermal or irradiative 'aging' of a particular silicone elastomer system. Despite these limitations, degradative pyrolysis-GC/MS coupled with PCA

http://dx.doi.org/10.5772/57509

has been shown to be a rapid and effective investigative and predictive tool for analysis of complex silicone elastomers.

Acknowledgements

Portions of this work were performed under the auspices of the U.S. Department of Energy by Lawrence Livermore National Laboratory under Contract DE-AC52-07NA27344. The authors also gratefully acknowledge Mark E. Pearson and Cynthia T. Alviso (LLNL) for their assistant with TGA and photography.

Author details

James P. Lewicki and Robert S. Maxwell

Lawrence Livermore National Laboratory, Livermore, CA, USA

References

[1] Arkles B. Look what you can make out of silicones. Chem Tech. 1983;13 (9):542-55. PubMed PMID: WOS:A1983RF44300007.

[2] Yoda R. Elastomers for biomedical applications. Journal of Biomaterials Science-Polymer Edition. 1998;9 (6):561-626. PubMed PMID: WOS:000074512300004.

[3] Allen KJ. Reel to real: Prospects for flexible displays. Proceedings of the Ieee. 2005 Aug;93 (8):1394-9. PubMed PMID: ISI:000230737600002.

[4] Yoon J, Baca AJ, Park SI, Elvikis P, Geddes JB, Li LF, et al. Ultrathin silicon solar microcells for semitransparent, mechanically flexible and microconcentrator module designs. Nature Materials. 2008 Nov;7 (11):907-15. PubMed PMID: WOS:000260472800025.

[5] Chinn SC, Alviso CT, Berman ESF, Harvey CA, Maxwell RS, Wilson TS, et al. MQ NMR and SPME Analysis of Nonlinearity in the Degradation of a Filled Silicone Elastomer. J Phys Chem B. 2010 Aug;114 (30):9729-36. PubMed PMID: WOS:000280361100002.

[6] Maxwell RS, Chinn SC, Alviso CT, Harvey CA, Giuliani JR, Wilson TS, et al. Quantification of radiation induced crosslinking in a commercial, toughened silicone rubber, TR55 by H-1 MQ-NMR. Polym Degrad Stab. 2009 Mar;94 (3):456-64. PubMed PMID: WOS:000264010500023.

[7] Wang SJ, Long CF, Wang XY, Li Q, Qi ZN. Synthesis and properties of silicone rubber organomontmorillonite hybrid nanocomposites. J Appl Polym Sci. 1998 Aug;69 (8):1557-61. PubMed PMID: WOS:000075150700010.

[8] Phillips SH, Haddad TS, Tomczak SJ. Developments in nanoscience: polyhedral silsesquioxane (POSS)-polymers oligomeric. Current Opinion in Solid State & Materials Science. 2004 Jan;8 (1):21-9. PubMed PMID: WOS:000222987600004.

[9] Liu C. Recent developments in polymer MEMS. Adv Mater. 2007 Nov;19 (22): 3783-90. PubMed PMID: WOS:000251383900003.

[10] Lewicki JP, Patel M, Morrell P, Liggat J, Murphy J, Pethrick R. The stability of polysiloxanes incorporating nano-scale physical property modifiers. Science and Technology of Advanced Materials. 2008;9 (2):024403.

[11] Pocknell D, Thomas DR, Kendrick TC. Reinforcement in silicone rubber. Rheological studies of reinforcing and non-reinforcing fillers in silicone fluids. Rubber and Plastics Age. 1969;50 (9):690-&. PubMed PMID: WOS:A1969E371600019.

[12] Traeger RK, Castongu.Tt. Effect of gamma-radiation on dynamic mechanical properties of styrene-butadiene rubbers. J Appl Polym Sci. 1966;10 (3):491-&. PubMed PMID: WOS:A19667468700013.

[13] Bajaj P, Babu GN, Khanna DN, Varshney SK. Room temperature-vulcanized silicone elastomer - effect of curing conditions and the nature of filler on mechanical and thermal-properties. J Appl Polym Sci. 1979;23 (12):3505-14. PubMed PMID: WOS:A1979GZ79600007.

[14] Hadjoudj A, David JC, Vergnaud JM. Model for predicting mechanical-properties of silicone sheet during pyrolysis with constant heating rate. Thermochim Acta. 1986 Jan;97:271-9. PubMed PMID: WOS:A1986AXN2400028.

[15] Tobolsky AV, Prettyman IB, Dillon JH. Stress relaxation of natural and synthetic rubber stocks. J Appl Phys. 1944 Apr;15 (4):380-95. PubMed PMID: WOS: 000200100400012.

[16] Lewicki JP, Liggat JJ, Pethrick RA, Patel M, Rhoney I. Investigating the ageing behavior of polysiloxane nanocomposites by degradative thermal analysis. Polym Degrad Stab. 2008 Jan;93 (1):158-68. PubMed PMID: WOS:000253305500020.

[17] Lewicki JP, Beavis PW, Robinson MWC, MAxwell RS. A Dielectric Relaxometry Study of Segmental Dynamics in PDMS/Boron Composite and Hybrid Elastomers. Polymer. submittted.

[18] Maxwell RS, Chinn SC, Solyom D, Cohenour R. Radiation-induced cross-linking in a silica-filled silicone elastomer as investigated by multiple quantum H-1 NMR. Macromolecules. 2005 Aug;38 (16):7026-32. PubMed PMID: WOS:000230978500040.

[19] Saalwachter K. Proton multiple-quantum NMR for the study of chain dynamics and structural constraints in polymeric soft materials. Prog Nucl Magn Reson Spectrosc. 2007 Aug;51 (1):1-35. PubMed PMID: ISI:000247284300001.

[20] Saalwachter K, Ziegler P, Spyckerelle O, Haidar B, Vidal A, Sommer JU. H-1 multiple-quantum nuclear magnetic resonance investigations of molecular order distributions in poly (dimethylsiloxane) networks: Evidence for a linear mixing law in bimodal systems. J Chem Phys. 2003 Aug;119 (6):3468-82. PubMed PMID: ISI: 000184350300056.

[21] Kimmich R, Fatkullin N. Polymer chain dynamics and NMR. Nmr - 3d Analysis - Photopolymerization. 2004;170:1-113. PubMed PMID: WOS:000227578100001.

[22] Spiess HW. Interplay of Structure and Dynamics in Macromolecular and Supramolecular Systems. Macromolecules. 2010 Jul;43 (13):5479-91. PubMed PMID: WOS: 000279573600001.

[23] Mayer BP, Lewicki JP, Weisgraber TH, Small W, Chinn SC, Maxwell RS. Linking Network Microstructure to Macroscopic Properties of Siloxane Elastomers Using Combined Nuclear Magnetic Resonance and Mesoscale Computational Modeling. Macromolecules. 2011 Oct;44 (20):8106-15. PubMed PMID: WOS:000295907200027.

[24] Provder T, Urban MW, Barth HG, editors. Differential scanning calorimetry—fourier transform IR spectroscopy and thermogravimetric analysis—fourier transform IR spectroscopy to differentiate between very similar polymer materials: American Chemical Society; 1994.

[25] Ettre K, Varadi PF. Pyrolysis-gas chromatographic technique for direct analysis of thermal degradation products of polymers. Anal Chem. 1962;34 (7):752-&. PubMed PMID: WOS:A19624029A00020.

[26] Ke B. Differential thermal analysis of high polymers.6. comments on some material and experimental factors. Journal of Polymer Science Part a-General Papers. 1963;1 (4):1453-&. PubMed PMID: WOS:A19636802B00017.

[27] Patel M, Chinn S, Maxwell RS, Wilson TS, Birdsell SA. Compression set in gas-blown condensation-cured polysiloxane elastomers. Polym Degrad Stab. 2010 Dec;95 (12): 2499-507. PubMed PMID: WOS:000285851800036.

[28] Grassie N, Macfarlane IG. Thermal-degradation of polysiloxanes.1. poly (dimethylsiloxane). Eur Polym J. 1978;14 (11):875-84. PubMed PMID: WOS:A1978FX16600001.

[29] Grassie N, Macfarlane IG, Francey KF. Thermal-degradation of polysiloxanes.2. poly (methylphenylsiloxane). Eur Polym J. 1979;15 (5):415-22. PubMed PMID: WOS:A1979GY96900001.

[30] Grassie N, Francey KF. The thermal-degradation of polysiloxanes.3. poly (dimethylmethyl phenyl siloxane). Polym Degrad Stab. 1980;2 (1):53-66. PubMed PMID: WOS:A1980KS80700004.

[31] Grassie N, Francey KF, Macfarlane IG. The thermal-degradation of polysiloxanes.4. poly (dimethyl-diphenyl siloxane). Polym Degrad Stab. 1980;2 (1):67-83. PubMed PMID: WOS:A1980KS80700005.

[32] Grassie N, Beattie SR. The thermal-degradation of polysiloxanes.8. poly (methyl-phenyl-para-silphenylene-siloxane). Polym Degrad Stab. 1984;9 (1):23-40. PubMed PMID: WOS:A1984TD79000003.

[33] Grassie N, Beattie SR. The thermal-degradation of polysiloxanes.7. mechanism of degradation of poly (tetramethyl-para-silphenylene siloxane) and copolymers with dimethylsiloxane. Polym Degrad Stab. 1984;8 (3):177-93. PubMed PMID: WOS:A1984TA17300004.

[34] Grassie N, Beattie SR. The thermal-degradation of polysiloxanes.5. synthesis, characterization and thermal-analysis of poly (tetramethyl-p-silphenylene siloxane) and copolymers with dimethylsiloxane. Polym Degrad Stab. 1984;7 (2):109-26. PubMed PMID: WOS:A1984SL20500004.

[35] Grassie N, Beattie SR. The thermal-degradation of polysiloxanes.6. products of degradation of poly (tetramethyl-para-silphenylene siloxane) and copolymers with dimethylsiloxane. Polym Degrad Stab. 1984;7 (4):231-50. PubMed PMID: WOS:A1984SV05800004.

[36] Patnode W, Wilcock DF. Methylpolysiloxanes. J Am Chem Soc. 1946;68 (3):358-63. PubMed PMID: WOS:A1946UB30600004.

[37] Thomas TH, Kendrick TC. Thermal analysis of polydimethylsiloxanes.i. thermal degradation in controlled atmospheres. Journal of Polymer Science Part a-2-Polymer Physics. 1969;7 (3PA2):537-&. PubMed PMID: WOS:A1969D456700008.

[38] Thomas TH, Kendrick TC. Thermal analysis of polysiloxanes.2. thermal vacuum degradation of polysiloxanes with different substituents on silicon and main siloxane chain. Journal of Polymer Science Part a-2-Polymer Physics. 1970;8 (10):1823-&. PubMed PMID: WOS:A1970H518800016.

[39] Camino G, Lomakin SM, Lageard M. Thermal polydimethylsiloxane degradation. Part 2. The degradation mechanisms. Polymer. 2002 Mar;43 (7):2011-5. PubMed PMID: WOS:000173952000005.

[40] Osthoff RC, Bueche AM, Grubb WT. Chemical stress relaxation of polydimethylsiloxane elastomers. J Am Chem Soc. 1954;76 (18):4659-63. PubMed PMID: WOS:A1954UB50700051.

[41] Lewis FM. The Science and Technology of Silicone Rubber. Rubber Chem Technol. 1962;35 (5):1222-75.

[42] Vanderweij FW. The action of tin-compounds in condensation-type rtv silicone rubbers. Makromolekulare Chemie-Macromolecular Chemistry and Physics. 1980;181 (12):2541-8. PubMed PMID: WOS:A1980KX11500011.

[43] Patel M, Skinner AR. Thermal ageing studies on room-temperature vulcanised polysiloxane rubbers. Polym Degrad Stab. 2001;73 (3):399-402. PubMed PMID: WOS:000170825400006.

[44] Patel M, Skinner AR. The effect of thermal aging on the non-network species in room temperature vulcanized polysiloxane rubbers. In: Clarson SJ, Fitzgerald JJ, Owen MJ, Smith D, VanDyke ME, editors. Synthesis and Properties of Silicones and Silicone-Modified Materials. Acs Symposium Series. 8382003. p. 138-50.

[45] Patel M, Soames M, Skinner AR, Stephens TS. Stress relaxation and thermogravimetric studies on room temperature vulcanised polysiloxane rubbers. Polym Degrad Stab. 2004 Jan;83 (1):111-6. PubMed PMID: WOS:000187855300014.

[46] Patel M, Skinner AR, Maxwell RS. Sensitivity of condensation cured polysiloxane rubbers to sealed and open-to air thermal ageing regimes. Polym Test. 2005 Aug;24 (5):663-8. PubMed PMID: WOS:000229979000019.

[47] Maiti A, Weisgraber T, Dinh LN, Gee RH, Wilson T, Chinn S, et al. Controlled manipulation of elastomers with radiation: Insights from multiquantum nuclear-magnetic-resonance data and mechanical measurements. Physical Review E. 2011 Mar;83 (3). PubMed PMID: WOS:000288698800003.

[48] Maxwell RS, Gee RH, Baumann T, Lacevic N, Herberg JL, Chinn SC. Characterization of Complex Engineering Silicones by H-1 Multiple Quantum NMR and Large Scale Molecular Dynamics Simulations. Advances in Silicones and Silicone-Modified Materials. 2010;1051:75-84. PubMed PMID: WOS:000305197500007.

[49] Chasse W, Lang M, Sommer JU, Saalwachter K. Cross-Link Density Estimation of PDMS Networks with Precise Consideration of Networks Defects. Macromolecules. 2012 Jan;45 (2):899-912. PubMed PMID: WOS:000299366300032.

[50] Wunderlich B. Thermal Analysis of Polymeric Materials. New York: Springer Publishing; 2005.

[51] Lewicki JP, Liggat JJ, Patel M. The thermal degradation behaviour of polydimethylsiloxane/montmorillonite nanocomposites. Polym Degrad Stab. 2009 Sep;94 (9): 1548-57. PubMed PMID: WOS:000269106300031.

[52] Lewicki JP, Mayer BP, Alviso CT, Maxwell RS. Thermal Degradation Behavior and Product Speciation in Model Poly (dimethylsiloxane) Networks. Journal of Inorganic and Organometallic Polymers and Materials. 2012 May;22 (3):636-45. PubMed PMID: WOS:000303417300011.

[53] Lewicki JP, Albo RLF, Alviso CT, Maxwell RS. Pyrolysis-gas chromatography/mass spectrometry for the forensic fingerprinting of silicone engineering elastomers. Journal of Analytical and Applied Pyrolysis. 2013 Jan;99:85-91. PubMed PMID: WOS: 000315240700012.

[54] Wachholz S, Keidel F, Just U, Geissler H, Kappler K. Analysis of a mixture of linear and cyclic siloxanes by cryo-gas chromatography fourier-transform infrared-spectro-

scopy and gas-chromatography mass-spectrometry. J Chromatogr A. 1995 Feb;693 (1):89-99. PubMed PMID: WOS:A1995QJ64300011.

[55] McNeill IC, Ackerman L, Gupta SN, Zulfiqar M, Zulfiqar S. Analysis of degradation products by thermal volatilization analysis at subambient temperatures. Journal of Polymer Science Part a-Polymer Chemistry. 1977;15 (10):2381-92. PubMed PMID: WOS:A1977DW34100008.

[56] C. MI. Thermal analysis of polymers. In: N. G, editor. Developments in Polymer Degradation. 1: Applied Science Publishers; 1984. p. 43-67.

[57] J. MW. Evolved gas analysis using vacuum pyrolysis. In: N. G, editor. Developments in Polymer Degradation. 5: Applied Science Publishers; 1977. p. 1-31.

[58] Allan D, Daly J, Liggat JJ. Thermal volatilisation analysis of TDI-based flexible polyurethane foam. Polym Degrad Stab. 2013 2//;98 (2):535-41.

[59] France LJ, Apperley DC, Ditzel EJ, Hargreaves JSJ, Lewicki JP, Liggat JJ, et al. An investigation of the nature and reactivity of the carbonaceous species deposited on mordenite by reaction with methanol. Catalysis Science & Technology. 2011;1 (6): 932-9. PubMed PMID: WOS:000294016600010.

[60] Lewicki JP, Pielichowski K, De la Croix PT, Janowski B, Todd D, Liggat JJ. Thermal degradation studies of polyurethane/POSS nanohybrid elastomers. Polym Degrad Stab. 2010 Jun;95 (6):1099-105. PubMed PMID: WOS:000278750800025.

[61] Lewicki JP, Liggat JJ, Hayward D, Pethrick RA, Patel M. Degradative Thermal Analysis and Dielectric Spectroscopy Studies of Aging in Polysiloxane Nanocomposites. In: Celina MC, Wiggins JS, Billingham NC, editors. Polymer Degradation and Performance. ACS Symposium Series. 10042009. p. 239-54.

[62] Kumooka Y. Analysis of deteriorated rubber-based pressure sensitive adhesive by pyrolysis-gas chromatography/mass spectrometry and attenuated total reflectance Fourier transform infrared spectrometry. Forensic Science International. 2006 Nov; 163 (1-2):132-7. PubMed PMID: WOS:000241463200014.

[63] K. V, P. F. Introduction to Multivariate Statistical Analysis in Chemometrics. Boca Raton: CRC Press; 2009.

[64] Hall AD, Patel M. Thermal stability of foamed polysiloxane rubbers: Headspace analysis using solid phase microextraction and analysis of solvent extractable material using conventional GC-MS. Polym Degrad Stab. 2006 Oct;91 (10):2532-9. PubMed PMID: WOS:000239536200035.

[65] Patel M, Morrell P, Cunningham J, Khan N, Maxwell RS, Chinn SC. Complexities associated with moisture in foamed polysiloxane composites. Polym Degrad Stab. 2008 Feb;93 (2):513-9. PubMed PMID: WOS:000254146200025.

Chapter 8

Measurement of Off-Gases in Wood Pellet Storage

Fahimeh Yazdanpanah, Shahab Sokhansanj, Hamid Rezaei, C. Jim Lim, Anthony K. Lau, X. Tony Bi, S. Melin, Jaya Shankar Tumuluru and Chang Soo Kim

Additional information is available at the end of the chapter

http://dx.doi.org/10.5772/58301

1. Introduction

Gas emissions (CO_2, CO, CH_4 and volatile organic compounds) from woody biomass storage systems have been studied extensively in recent years. The decomposition is due to biological, chemical oxidative and slow pyrolytic processes depending on the availability of oxygen [1-3]. The major VOCs emitted from stored wood pellets were aldehydes, some of which are known to cause irritation to the respiratory system [4].

Previous measurements have shown that freshly made wood pellets continue to emit flammable gases CO, H_2, and CH_4 during storage and handling. A large portion of this emission is CO as opposed to emission of volatile organic compounds (VOC) like formic acid, methanol and aldehydes [5, 6]. More volatile organic compounds emit from stored sawdust and wood chips compared to stored wood pellets [7-9]. This is because of the fact that most of the VOCs easily evaporate during heating and drying processes [10, 11] and as wood pellets pass through high temperature during drying and pelletization, decomposition of extractives would result in less VOCs emission. During pelletization, sawdust is dried at temperatures from 100 to 400 °C so the moisture content goes below 10%. Roffael [12], Rupar and Sanati [13] also reported the emission of VOCs during biomass storage. Large amount of monoterpenes and other volatile organic compounds are emitted during drying [10, 11, 14, 15]. High levels of CO were first measured in compartments of a ship in 2002 when an ocean vessel was discharging wood pellets in Rotterdam, The Netherlands. The wood pellets were shipped from Vancouver, British Columbia and the duration of the voyage was 38 days. One person was killed and several people were severely injured as a result of exposure to high concentration of CO when entering a stairway adjacent to the cargo space [16]. Svedberg identified that the storage of wood pellets led to the emission of high levels of hexanal as a result of the general degradation processes of wood.

In terms of CO_2, CO, and CH_4 emissions, higher emission factors are associated with higher temperatures, whereas increased relative humidity in the enclosed container increased the rate of gas emission and a corresponding depletion of oxygen [17, 18]. Wihersaari [8] concluded that the CO_2 emissions from fresh forest residues were almost three times higher than the dried materials, and suggested that mixing the heaps during the storage period would probably cause increased emissions rates. Emissions of CO, CO_2 and CH_4 are likely due to biodegradation and auto-oxidative reactions of organic constituents naturally present in wood [1].

2. Experimental set-up

2.1. Small scale experiments

The off-gassing tests were conducted at three nominal temperatures of 25 °C, 40 °C and 60 °C. Room temperature provided the 25 °C storage environment. Two identical ovens were used to provide 40 °C and 60 °C storage environments. Twenty four (24) glass containers, 2L each, were divided into three equal groups for testing at the three temperatures. For each set of 8 containers, 6 containers were used for gas sampling and 2 containers were used for temperature measurement. The temperature inside the containers was measured with thermocouples and data was logged onto a PC. The following procedure was followed for each test container. The empty weight of the container without the lid was recorded (M_1). The container was filled to 75% volume with wood pellets (M_2). Finally, the weight of the wood pellets was calculated (M_3). The lid with sampling port was then applied, thereby sealing the container. Filled containers along two empty containers (for each storage temperature) were arranged in upright position. The two empty jars were included to have consistent condition as the filled jars for temperature reading. Temperature was also directly recorded from a thermocouple inside the oven.

2.2. Pilot scale experiments

All measurements of the gases (CO, CO_2 and CH_4) were made in an experimental silo designed and installed at the Clean Energy Research Centre (CERC), the Department of Chemical & Biological Engineering, University of British Columbia [19, 20]. The experimental silo has a volume of 5.65m^3 (1.2m diameter, 4.6m height) and is made of mild steel. The silo is designed to represent the storage of substantial quantities of wood pellets. A two level structure was designed and built to support the pilot scale silo. The set-up was further equipped with sensors for temperature, gas pressure and relative humidity measurements. Thermocouples and pressure transducers were installed at various levels along the height of the silo. Two EXP-32 cards were used to log the temperature and pressure data during the experiments. The experimental data of pressure and temperature were logged into a computer running on the

http://dx.doi.org/10.5772/58301

LABTECH software using a data acquisition board (PCI-DAS08, Techmatron Instruments Inc, Canada) consisting of pressure transducers and thermocouples. However the pilot size of the set-up made it challenging to have it equipped with different monitoring systems and make it sealed. Fourteen 12mm diameter gas sampling ports with 3mm ball valves are provided at 2 sides of the pilot silo along the wall at different levels for gas sampling. The positions of the gas sampling ports in the large silo are indicated in Table 1. Six of the gas sampling ports are located on one side of the silo and another 7 on the other side.

Gas sampling ports on side A		Gas sampling ports on side B	
Gas Port Name	**Elevation (m)**	**Gas Port Name**	**Elevation (m)**
G0	4.16	G0	4.16
G2	3.55	G3	3.35
G4	2.94	G5	2.74
G6	2.33	G7	2.13
G8	1.65	G9	1.52
G10	1.21	G11	0.91
G12	0.50	G13	0.30

Table 1. Location of the gas sampling ports on the experimental silo

3. Materials and methods

3.1. Small and pilot scale experiments

In small scale tests, initially and at intervals, approximately 25 mL of gas was drawn from each container by an air-tight GC syringe (25mL SGE Gas-Tight Syringe, Luer-Lock and TOGAS Luer Lock Adapter, Mandel Scientific Company) and analyzed by GC/FID (Flame Ionization Detector) and GC/TCD (Thermal Conductivity Detector) methods for the composition of the sampled gases (CO, CO_2, CH_4, N_2, O_2 and H_2). In pilot scale tests, Gas samples were collected using an airtight syringe (SG-009770-100mL SGE Gas-Tight Syringe, Luer-Lock, Mandel Scientific, Canada). The syringe had a Luer lock device to help in collecting a known quantity of gas sample. The Shimadzu GC needle used with the 100mL SGE syringe was Togas Loop Fill Interface N711 Needle (Model number: 220-90615-00, Mandel Scientific Inc.).The GC has three columns in series: Porapak-N (80/100 mesh, 3 m), Porapak-Q (80/100 mesh, 3 m) and a MS-5A (60/80, 2.25 m). The FID detector was used for CO, CO_2 and CH_4 and the TCD was used for N_2, O_2 and H_2, with Argon as the carrier gas. Argon was the reference gas for the TCD. Compressed air was the reference gas for the FID.

Gas sampling started one day after loading of the containers. Prior to gas concentration measurements, the GC was calibrated with three different standard gases, which contained known concentrations of carbon monoxide (CO), carbon dioxide (CO_2), methane (CH_4), oxygen (O_2), nitrogen (N_2), hydrogen (H_2) and helium (He) (Table 2). To ensure steady and accurate readings by the GC, 25 mL of a standard gas with known concentration of gases was injected to the GC for calibration before and after gas analysis every day.

Standard gas 1		**Standard gas 2**		**Standard gas 3**	
Components	Concentration (%)	Components	Concentration (%)	Components	Concentration (%)
CO	2.5	CO	0.05	CO	0.50
CO_2	6.0	CO_2	0.10	CO_2	0.50
CH_4	1.5	CH_4	0.50	CH_4	0.10
N_2	3.0	H_2	0.50	H_2	1.00
O_2	10.0	Air	Balance	Air	balance
He	2.0				
Ar	balance				

Table 2. Compositions (volumetric) of standard gases for GC calibration

4. Results and discussion

4.1. Small scale tests

4.1.1. Off-gas concentration at high temperature and high moisture

The moisture content of the reference wood pellets was 4% as received. They were subsequently conditioned to 9, 15, 35 and 50% moisture content respectively by spraying distilled water on the wood pellets in a container during rotary motion. At 15% moisture or higher the integrity and wood pellet shape were compromised.

The off-gassing tests were conducted at three temperatures 25°C, 40°C and 60°C. For the tests at 25°C, the containers were placed on the lab shelf. To provide the higher storage temperatures two identical ovens were used. The Biomass and Bioenergy Research Group (BBRG) at UBC has developed the method of using multiple smaller containers and sampling a very limited amount of gas from each container at lower frequency rather than frequent sampling from a larger container in order to minimize the distortion in gas concentration due to extraction of gas. This method allows a much smaller sample size to be used for testing.

http://dx.doi.org/10.5772/58301

Eight glass containers with wood pellets were put on the lab shelf for 25°C for testing. The 8 containers were named A-4-X, A-4-Y, A-9-X, A-9-Y, A-15-X and A-15-Y [e.g. A-4-X indicates the first replicate of wood pellets sample at ambient temperature (25°C) and 4% moisture content]. The same format was used for labeling the samples at higher temperature. The letter A was replaced by 40 for the tests at 40°C and with 60 for the tests at 60°C. Each set of 8 containers (6 containers for gas sample measurement and 2 for temperature measurement) were placed in 40°C and 60°C ovens. The temperature inside the containers was measured with thermocouple sensor connected to a computer for temperature logging. All containers were checked before conducting the experiments to ensure proper sealing. The exact same labeling and procedure were used for tests at 15, 35 and 50% moisture content.

To create the baseline for gas concentrations, two empty containers were placed in room temperature (25°C), another two in the 40°C oven and another two in the 60°C oven. 100 mL of gas was drawn from each container and analyzed by the GC for gas composition. Figure 1 to Figure 10 show plots of gas concentrations vs. storage time for the three temperatures and 2 moisture concentrations [4 and 50%] for each gas. The concentration of each gas is presented on the same graph for one moisture content and 3 different storage temperatures. The concentrations are given in percent on volumetric scale in ppmv (parts per million volume). A concentration of 0.6% means 6000 ppmv.

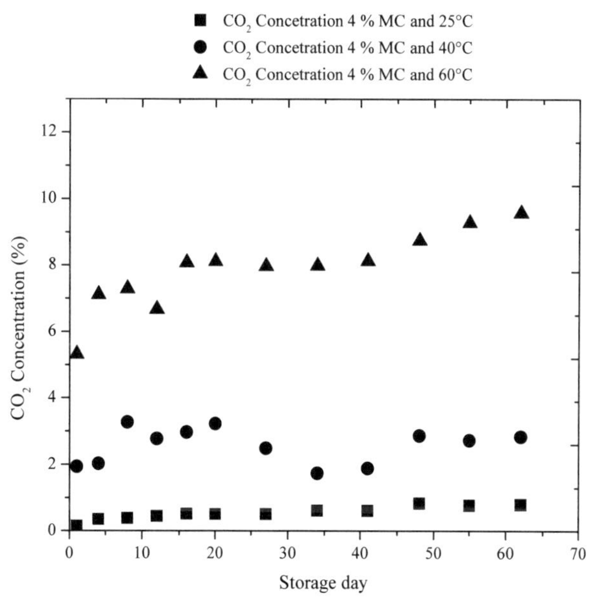

Figure 1. Concentration of CO_2 for wood pellet with 4% moisture content over 62 days at 25°C, 40°C and 60°C.

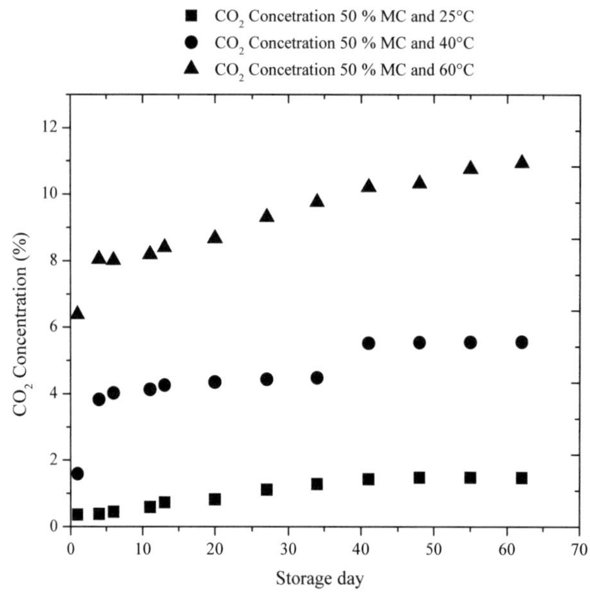

Figure 2. Concentration of CO_2 for wood pellet with 50% moisture content over 62 days at 25°C, 40°C and 60°C

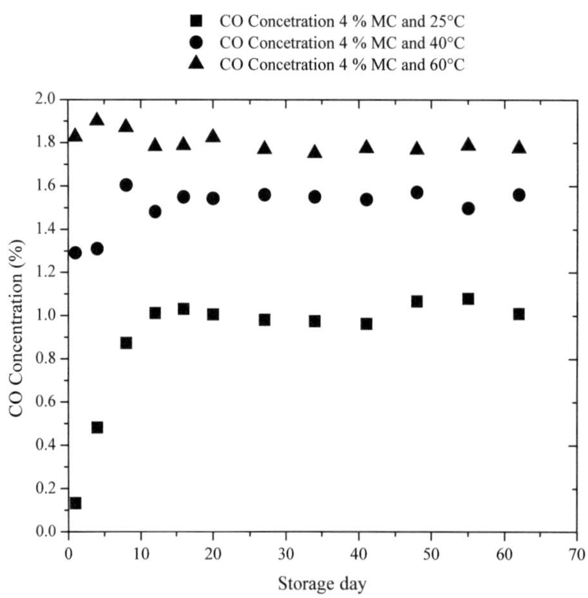

Figure 3. Concentration of CO for wood pellet with 4% moisture content over 62 days at 25°C, 40°C and 60°C

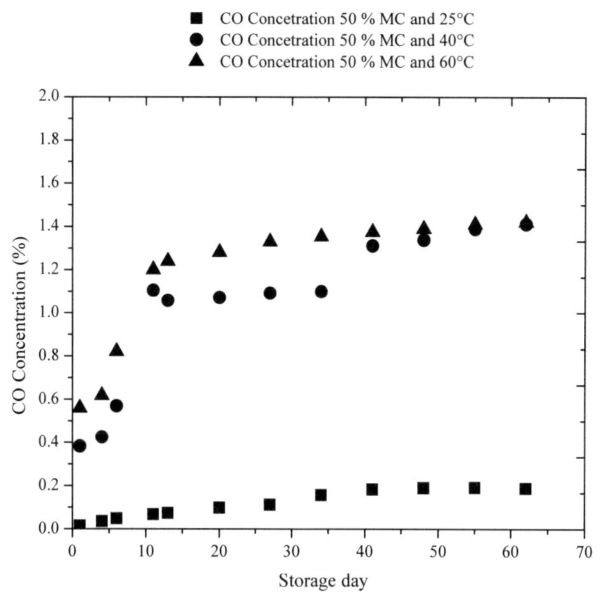

Figure 4. Concentration of CO for wood pellet with 50% moisture content over 62 days at 25°C, 40°C and 60°C

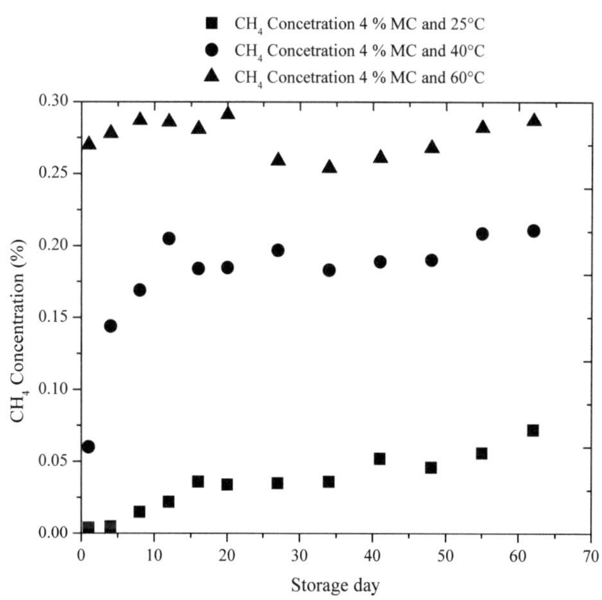

Figure 5. Concentration of CH_4 for emissions from wood pellet with 4% moisture content over 62 days at 25°C, 40°C and 60°C

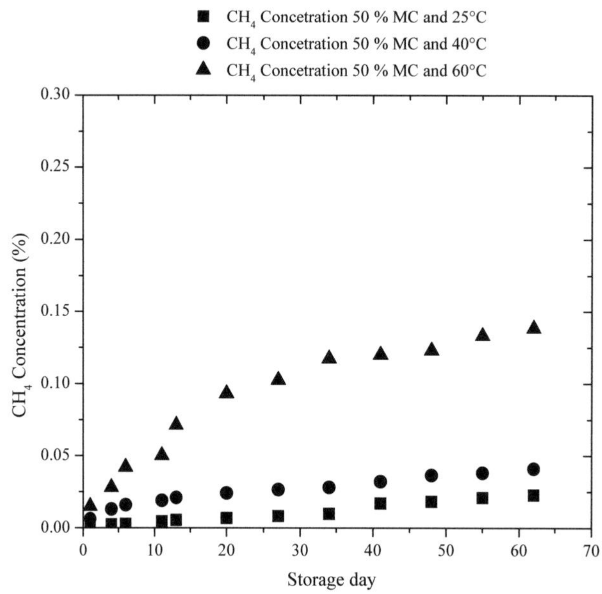

Figure 6. Concentration of CH_4 for emissions from wood pellet with 50% moisture content over 62 days at 25°C, 40°C and 60°C

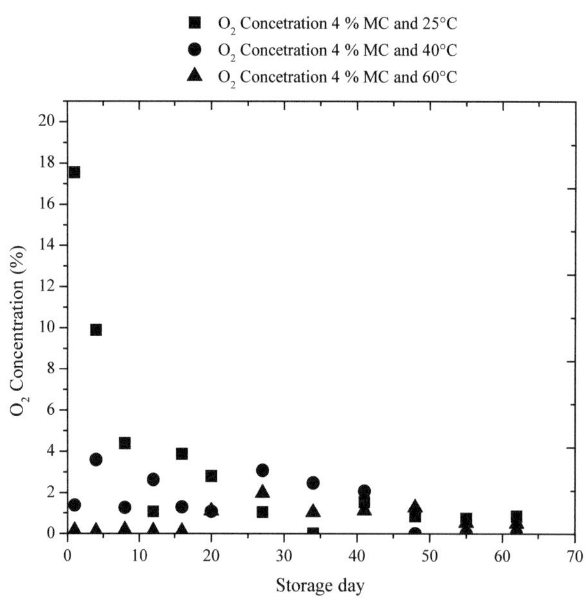

Figure 7. Oxygen depletion for wood pellet with 4% moisture content over 62 days at 25°C, 40°C and 60°C

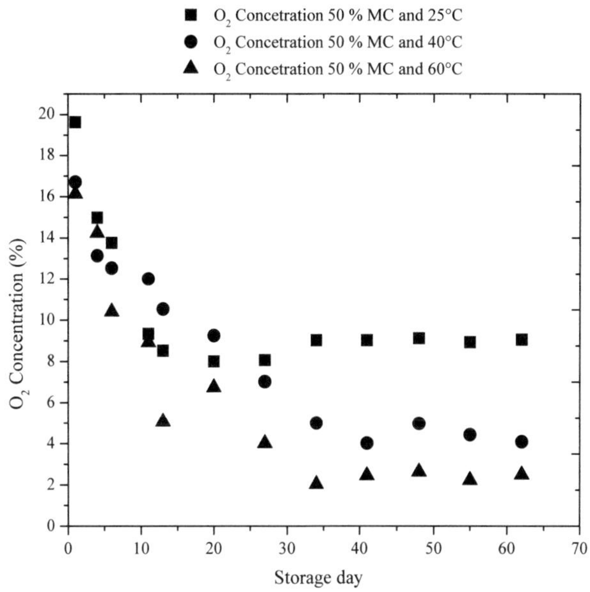

Figure 8. Oxygen depletion for wood pellet with 50% moisture content over 62 days at 25°C, 40°C and 60°C

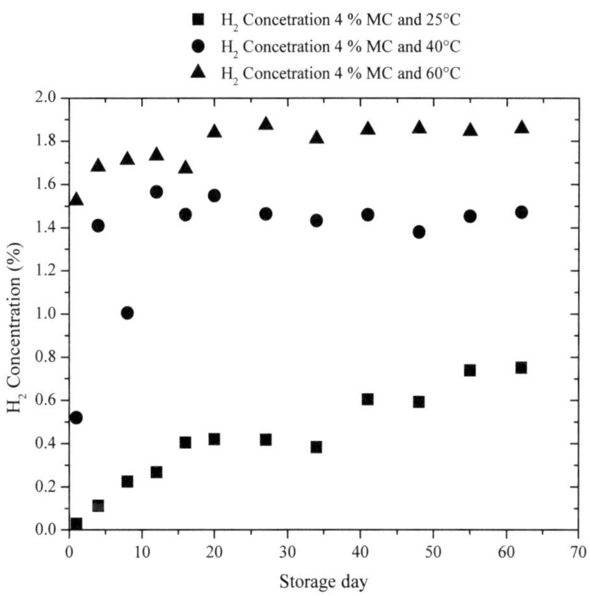

Figure 9. Concentration of hydrogen for emissions from wood pellet with 4% moisture content over 62 days at 25°C, 40°C and 60°C

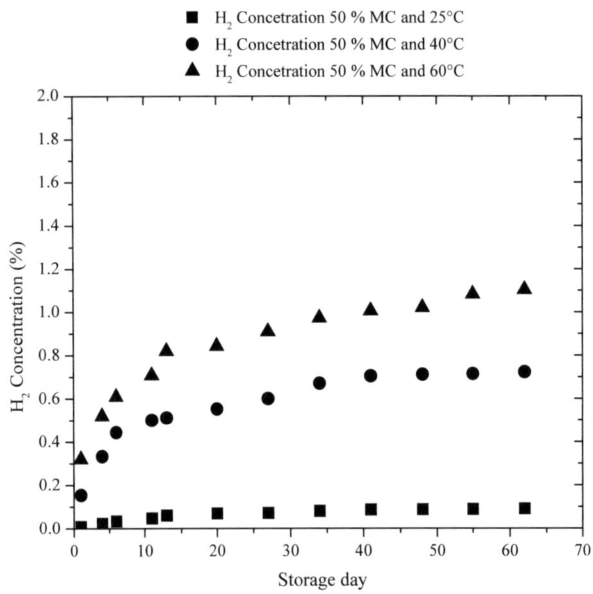

Figure 10. Concentration of hydrogen for emissions from wood pellet with 50% moisture content over 62 days at 25°C, 40°C and 60°C

Plots in Figure 1 to 2 show the concentration of carbon dioxide over time at 25°C, 40°C and 60°C for wood pellets with 4 and 50% moisture content respectively. Figures of carbon dioxide concentration for wood pellets with 9, 15 and 35% moisture content are presented in Appendix A. As expected, the higher the temperature, the higher the concentration of the CO_2 gas. However, for wood pellets with 4, 9 and 15% moisture content at 25°C, the CO_2 concentration increased from ~0.9% to ~1.7% instead showing a relation between time and moisture content. At 25°C when the moisture content increased higher to 35 and 50%, the maximum concentration of CO_2 did not go beyond what was seen for the wood pellets with 15% moisture content. At 40°C and 60°C, the increase was from 2.84% to 5.55% and 9.58% to 10.94% respectively (for wood pellets with 4% to 50%). The maximum concentration of CO_2 was observed at 15% moisture content. Above 15% moisture content, the maximum concentration of CO_2 increased only marginally at 40°C and stayed almost the same at 60°C.

Figure 3 to 4 show CO concentration at 4 and 50% moisture content at 25°C, 40°C and 60°C [Rest of the Figures for CO can be found in Appendix A]. The concentration of CO increased over time for wood pellets with 4 and 9% moisture content as the storage temperature increased. CO concentration did not increase at 25°C, 40°C and 60°C for wood pellets with 15% moisture content. The CO concentrations at room temperature and the six moisture contents the peak concentration of CO was the same for 4% and 9% moisture content and decreased to ~0.8% for wood pellets with 15% moisture content. The concentration decreased even further for wood pellets with 35 and 50% moisture content. At 40°C, for wood pellets with 4 and 9 and 15%, carbon monoxide concentration decreased from 1.5% to 1.4% and 1.1% respectively. For wood pellets with 35%, the peak concentration stayed at 1.0% and increased slightly when the moisture content of wood pellets increased to 50%. The same CO concentration change was observed at 60°C for wood pellets with 4 and 9 and 15%. When the moisture content increased to 35%, CO peak concentration increased to 1.52% but stayed the same for higher moisture content like 50%.

Similar to carbon dioxide concentration, methane concentration over time increased with storage temperature for all levels of moisture contents except for 35% where the concentration was quite the same at 40°C and 60°C. An increase in temperature from 25°C to 40°C caused the peak concentration of CO to increase from 0.07% to 0.21%. Further increase in temperature to 60°C, resulted in peak concentration of CO to 0.28%. Except for the gas concentration at 25°C, concentration of CO decreased over time as the moisture content increased to 40°C and 60°C respectively.

Plots in Figure 7 to 8 as well as Figure A.10 to A.12 show the concentration of oxygen for wood pellets with six different moisture content at 25°C followed by the concentration at higher temperatures (40°C and 60°C) as a function of time. Depletion of oxygen was least significant for samples at 25°C and moisture content higher than 15%. For wood pellets with 4%, 15%, 35% and 50% moisture content the oxygen content decreased as a function of time. However, for wood pellets with 9% moisture content the oxygen depletion was more rapid during the first few days, at 25°C to 40°C. After 20 days the, oxygen concentration was leveling out at 5-6% at 25°C and reached 0% very quickly at 40°C.

Figure 9 to 10 shows the concentration of hydrogen for wood pellets with 4 and 50% moisture content over time at 25°C followed by the concentration at higher temperatures (40°C and 60°C). The release of hydrogen was minimal for samples with 50% moisture at 25°C. It was concluded that for the same moisture content, the hydrogen concentration increased as the temperature increased. However for wood pellets at the same temperature, the hydrogen concentration decreased as the moisture content increased.

4.1.2. Estimation of flammability of emissions from wood pellets

This research conducted to provide a scientific basis to determine whether or not the concentration of the composite mix of gases emitted from wood pellets could reach flammable concentrations and if so, under what conditions. Table 3 summarizes the results of the calculations using the ISO 10156:2010 Standard [21].

The $\sum_{i=1}^{n} A_i[\frac{100}{T_{Ci}} - 1]$ factor in
Table 4 represents the equivalent concentration of flammable gases as measured.

The $\sum_{k=1}^{p} B_k K_k$ factor in Table 4 represents the equivalent concentration of the flammable gases in a mixture of nitrogen as calculated, which if mixed with full complement of air (oxygen), may sustain a flame if ignited with an external source. Since the oxygen content is extremely low in an enclosed cargo hold there is even lower risk of fires and explosions in the oxygen deprived conditions in an enclosed cargo hold than in open atmospheric conditions saturated with oxygen.

Wood pellets exported to Europe have typically between 4 and 6% moisture. The 15% and higher represented wood pellets which have been accidentally exposed to moist conditions for a prolonged period of time. Testing of wood pellets with abnormal moisture content is important to make sure there is good safety margin for long duration shipments such as a 30-34 days voyage from for example British Columbia to Europe. The emission of gases from wood pellets reached a peak (plateau) as summarized in Table 3 within the first 20-30 days. At the same time the oxygen content decreased rapidly to a very low level as the emission of CO_2 increased resulting in a near inert condition in the space. The flammability of the mixed gases was calculated when exposed to air with full complement of oxygen as per the ISO Standard.The research was carried out over a period of 62 days, just to ensure the gas evolution continued to be stable. The same calculations are done for 62 days as well and presented in Table 4.

	25°C			40°C			60°C		
Gas Species	**4% Moisture**	**9% Moisture**	**15% Moisture**	**4% Moisture**	**9% Moisture**	**15% Moisture**	**4% Moisture**	**9% Moisture**	**15% Moisture**
CO_2	0.617	1.334	1.110	1.741	3.572	3.620	7.980	8.490	10.290
CO	0.975	1.119	0.678	1.551	1.329	1.120	1.752	1.551	1.120
CH_4	0.036	0.041	0.019	0.183	0.174	0.096	0.254	0.225	0.120
H_2	0.383	0.421	0.365	1.460	1.209	0.946	1.811	1.473	1.230
$CO/CH_4/H_2$ mixture (sum)	1.394	1.581	1.062	3.194	2.712	2.162	3.817	3.248	2.470
Sum	2.011	2.915	2.172	4.935	6.284	5.782	11.797	11.738	12.760
O_2	0.000	0.000	0.000	0.000	0.000	0.000	0.000	0.000	0.000
N_2	97.989	97.085	97.828	95.065	93.716	94.218	88.203	88.262	87.240
Flammability determination in accordance with ISO 10156 Standard									
$\sum_{i=1}^{n} A_i[\frac{100}{T_{Ci}}-1]$	12.398	13.907	10.253	35.659	30.013	23.510	43.556	36.311	28.641
$\sum_{k=1}^{p} B_k K_k$	98.915	99.086	99.493	97.677	99.074	99.648	100.173	100.997	102.675
Flammable	No	No	No	No	No	No	No	No	No

Table 3. Concentration of gases from wood pellets after 34 days of storage

http://dx.doi.org/10.5772/58301

Gas Species	25°C			40°C			60°C		
	4% Moisture	9% Moisture	15% Moisture	4% Moisture	9% Moisture	15% Moisture	4% Moisture	9% Moisture	15% Moisture
CO_2	0.805	1.643	1.593	2.840	4.680	4.820	9.571	10.925	11.890
CO	1.010	1.103	0.758	1.561	1.412	1.112	1.775	1.560	1.128
CH_4	0.072	0.086	0.062	0.211	0.177	0.114	0.287	0.241	0.132
H_2	0.753	0.680	0.559	1.472	1.251	1.071	1.858	1.482	1.261
$CO/CH_4/H_2$ mixture (sum)	1.835	1.869	1.379	3.244	2.840	2.297	3.920	3.283	2.521
Sum	2.640	3.512	2.972	6.084	7.520	7.117	13.491	14.208	14.411
O_2	0.000	0.000	0.000	0.000	0.000	0.000	0.000	0.000	0.000
N_2	97.360	96.488	97.028	93.916	92.480	92.883	86.509	85.792	85.589
Flammability determination in accordance with ISO 10156 Standard									
$\sum_{i=1}^{n} A_i[\frac{100}{T_{Ci}} - 1]$	19.328	18.740	14.484	36.215	31.229	25.802	44.838	36.702	29.345
$\sum_{k=1}^{p} B_k K_k$	98.568	98.953	99.418	98.176	99.500	100.113	100.866	102.179	103.424
Flammable	No	No	No	No	No	No	No	No	No

Table 4. Concentration of gases from wood pellets after 62 days of storage

5. Pilot scale tests

5.1. Off-gas concentration in silo head-space

Gas composition analysis was performed for the first 9 weeks after loading the pilot silo. Results (Figure 11 to Figure 13) showed a rapid increase in the concentration of CO_2, CO and CH_4 in the head-space of the silo. Off-gas concentrations in the first 7 days were 1.0-1.6% CO_2, 0.8-1.0% CO and 0.73- 0.85% CH_4. All gas concentrations increased with storage time. Figure 14 shows the increase in the CO/CO_2 ratio with time, which reached a constant value after 50 days of storage. Some scattering of data is seen between days 10 to 15 with a peak CO/CO_2 ratio of about 0.59 on day 10. Minimum oxygen concentration in the bed was also seen on day 10 (Figure 15). It could be explained by slightly (~2%) lower local relative humidity recorded by the cable and higher H_2 and thus the shift of water-gas shift reaction towards reactants. Oxygen concentrations were also measured according to the same gas sampling schedule. As shown in Figure 15 the O_2 level dropped dramatically from 21% to 7-8% after 3 days and reached close to 0% after 1 week and again after 3 weeks of storage. The concentration of carbon dioxide in the head-space increased to about 2.7% (27,000 ppm), and this is higher compared to the CO_2 concentrations in the pellet bed which could be attributed to the exposure of pellet surface to head-space oxygen and thus higher local emission. Lab-scale experiments had been previously performed to determine the effect of oxygen availability on the emission rates of off-gasses. Highest emission of CO_2 was seen in containers with higher percentage of head-space or where oxygen was pumped in regularly. When wood pellets were stored in helium, minimum emission of CO_2 was detected. For carbon monoxide, its concentration increased from the very first days of storage and reached a maximum value of about 1.7% after 9 weeks of storage. This accumulated concentration was well above the threshold [22] limit value (TLV) for human health, and it can cause injuries and immediate death.

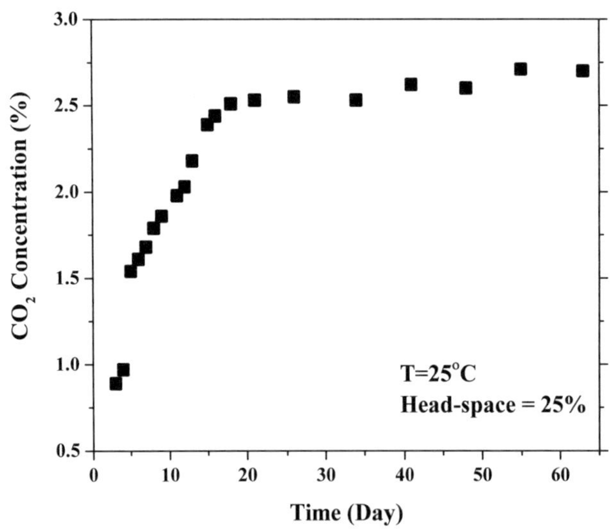

Figure 11. CO_2 concentration in head-space of the silo as a function of storage time for wood pellets. (G0: head-space gas sampling port) [Concentrations are volume percentage]

http://dx.doi.org/10.5772/58301

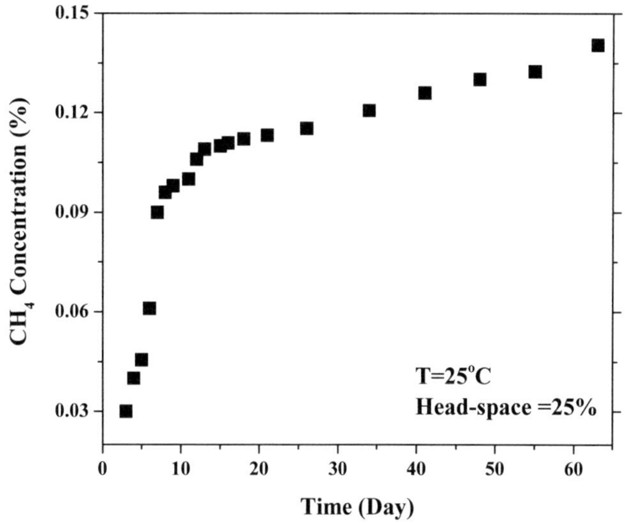

Figure 12. CH_4 concentration in head-space of the silo as a function of storage time for wood pellets. (G0 head-space gas sampling port) [Concentrations are volume percentage]

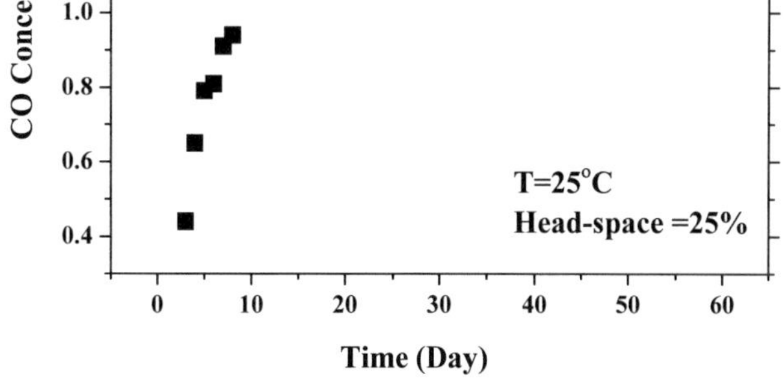

Figure 13. CO concentration in head-space of the silo as a function of storage time for wood pellets. (G0 head-space gas sampling port) [Concentrations are volume percentage]

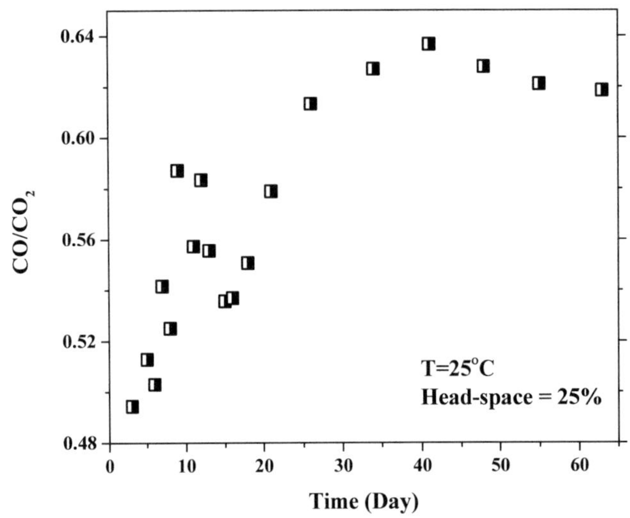

Figure 14. CO/CO_2 ratio in head-space of the silo as a function of storage time for wood pellets. (G0 head-space gas sampling port) [Concentrations are volume percentage]

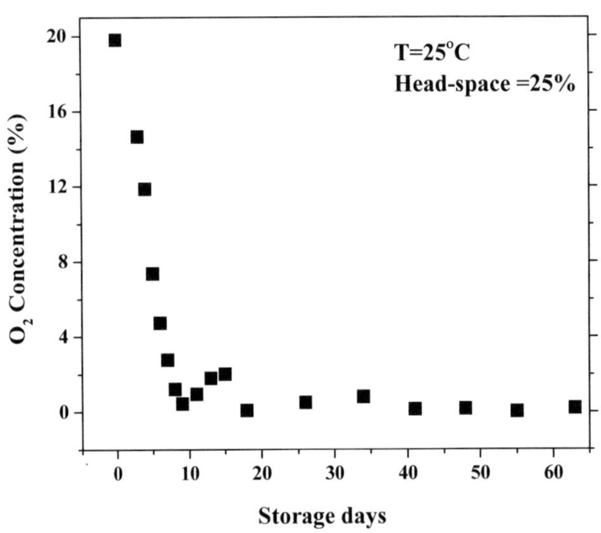

Figure 15. O_2 concentration in head-space of the silo as a function of storage time for wood pellets. (G0 head-space gas sampling port) [Concentrations are volume percentage]

http://dx.doi.org/10.5772/58301

The gas emissions are converted to emission factors using Eqn (1) in order to compare the results with other researchers' work on the same basis:

$$f_i = \frac{P(C_i V_g) M_{wt} C_{N0}}{RTM_P C_{Nt}} \quad (1)$$

where f_i is emission factor of species "i", C_i is the gas concentration, M_{wt} is the molecular weight (g),V_g is the gas volume (m^3), R is universal gas constant (8.314 J/mol K), T is temperature (K),P is the pressure (Pa), M_P is the mass of pellets (kg). Assuming that the amount of N_2 remains constant during off-gassing process, C_{N0}/C_{Nt} is introduced as a correction factor to balance the change in gas volume where C_{Nt} is the concentration of nitrogen measured at any time and C_{N0} is the concentration of nitrogen at the beginning of test. The gas emission results for 25% head-space are compared to those obtained by Kuang et al. [1] for the storage of wood pellets. As shown in Table 5, the peak emission factors of all three gases derived from this study (using a large pilot silo) are higher than their findings (using a small lab-scale reactor); though both sets of values are in the same order of magnitude.

Peak emission factor (g/kg)	**This study (Storage volume=5200 L)**	**Kuang et al. [1] (Storage volume=45 L)**
CO	0.0191	0.0124
CO_2	0.0485	0.0200
CH_4	0.0009	0.0002

Table 5. Comparison of peak emission factor for CO, CO_2 and $CH_4$with previous research

Methane emission is due to the activities of anaerobic microorganisms. For instance, in a typical anaerobic digestion process for biogas production from readily biodegradable feedstock, the methane content can range from 50-70%. In this study, the CH_4 concentration was observed to increase from 0.03% to 0.14% after 9 weeks, while an oxygen deficient environment was created in the silo just a few days after the start of the test. The relatively low CH_4 contents are in line with the results of total bacteria counts, which were less than 5 cfu/g of pellets as compared to orders of magnitude higher microbial counts in a typical anaerobic digester.

5.2. 3D analysis of longitudinal distribution of emitted gases

The measured concentrations of the gases derived from all 13 gas sampling ports over time are plotted in 3D graphs and contours. The contour plots would provide a supplemental view of how the various gases are stratified in the bed. Figure 16 show the 3D and contour plots of the concentration of carbon dioxide in all locations. For the first several days, the gas concentration was quite the same for all locations. After 10 days, CO_2 concentrations were different; yet, after about 40 days, the concentrations became uniform again everywhere in the silo. The contour plot clearly illustrates that the CO_2 concentration was always higher in the head-space compared to the other areas within the pellet bulk. CO_2 emission is more sensitive to temperature versus CO; this could explain why higher CO_2 concentration was observed in the silo head-space. While the peak CO_2 concentration of 2.7% was reached just after 2 weeks in the head-space of silo, such a high concentration was not seen in the pellet bulk section until after 45 days. Carbon dioxide concentration was above the threshold limit value for worker safety and this limit was reached sooner in the silo head-space. In terms of worker safety, both the Occupational Safety and Health Administration (OSHA) and the American Conference of Governmental Industrial Hygienists (ACGIH) have set the permissible exposure limit (PEL) and the threshold limit value (TLV), respectively for CO_2 of 5,000 ppm (0.5%) by volume over an 8-hour average exposure.

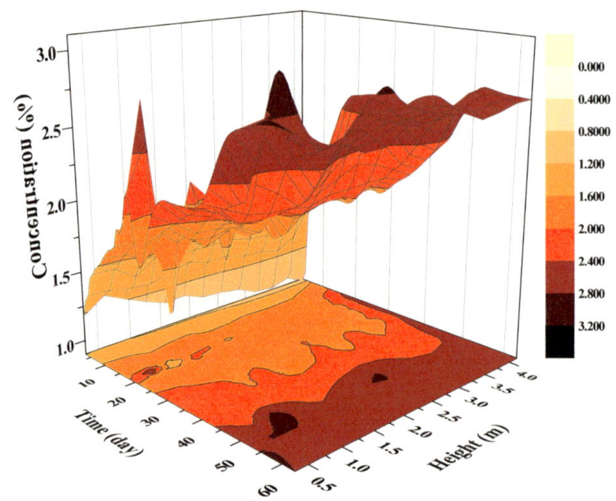

Figure 16. map of CO_2 concentration in the silo at all locations (G0 to G13) during 63 days of storage [Concentrations are volume percentage]

Figure 17 shows the overview of carbon monoxide emission, accumulation and dispersion over 63 days of storage. As distinguished from the concentration profile for carbon dioxide, a higher CO concentration was observed in the head-space of the silo at the beginning of the test, and the CO concentrations were similar in most other locations within the pellet bulk. This difference may be due to the gas dispersion phenomenon. Nevertheless, after 18 days of storage, stratification was observed with the highest concentration of CO at the G5, G6 and G7 locations (2.03-2.54 m from the bottom of the silo). Due to close value of density for air and carbon monoxide, not much gravitational stratification of CO was expected. Although carbon monoxide emissions are attributed to oxidation of unsaturated acids in wood pellets, the exact pathway through which CO is emitted hasn't been identified yet. More distinct stratification of gas was seen in this study for carbon monoxide compared to other off-gases, which could be attributed to the high uptake of CO by pellets during storage.

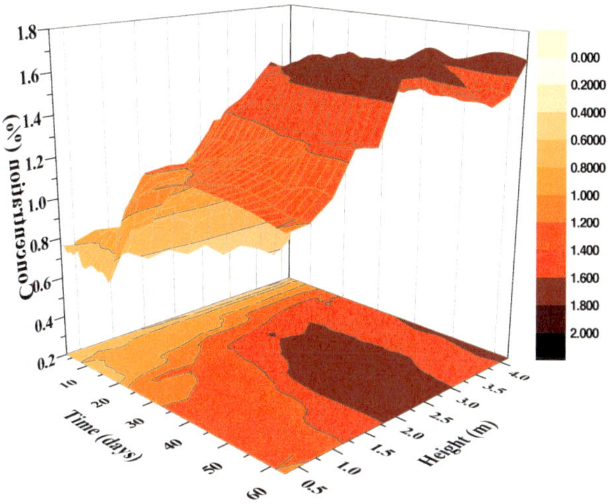

Figure 17. map of CO concentration in the silo at all locations (G0 to G13) during 63 days of storage [Concentrations are volume percentage]

Figure 18 illustrate the distribution of emitted methane (CH_4) during 63 days of sealed storage of wood pellets. After 4 days, the concentration of methane was the highest close to the bottom of the silo. Over time the emitted gas started to stratify; but after 48 days almost the same concentration was observed in all locations, with slightly higher CH_4 concentration at the upper sections of the silo near the interface of wood pellets and head-space. The same stratification was noted until the end of the test. Methane emission has a similar behavior as carbon dioxide; both gases reached higher concentrations more quickly in the head-space than in the other areas within the pellet bed.

Diffusion of gases in air and its effect on oxygen deficiency could also partly account for the created environment. Studies [23, 24] have been conducted to understand how easily the released gases (carbon dioxide, helium and sulfurhexafluoride) would mix with air and whether they would remain fully mixed. The gases were found to diffuse in air more readily than expected and modest gas velocities due to natural convection would fully mix the released gases with fresh air. In the wood pellet storage, gases were emitted over time from the first day till the gas concentrations reached a plateau. Some gas stratification was detected in the early weeks of storage especially for carbon monoxide which is due to chemical and physical adsorption between the material and the gases as well as differences in temperature and relative humidity at various levels in the silo. However after a few weeks, gases started to mix and remained mixed. In this regard, natural convection within the silo can bring about gas mixing.

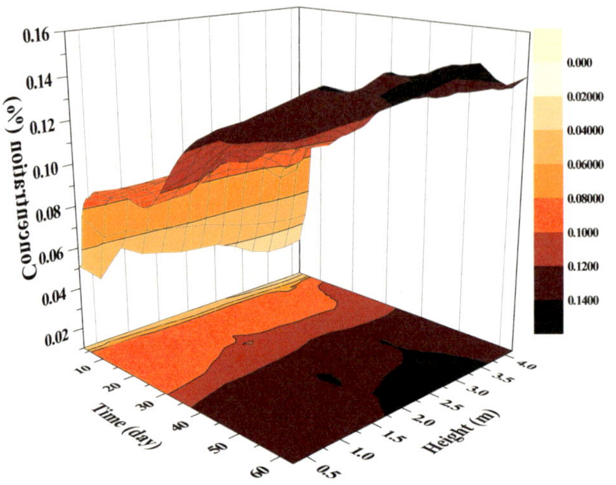

Figure 18. map of CH_4 concentration in the silo at all locations (G0 to G13) during 63 days of storage [Concentrations are volume percentage]

5.3. Oxygen depletion

Figure 19 demonstrates the oxygen depletion in the silo head-space during storage of wood pellet, which is much faster than the oxygen within the bed of wood pellets. Oxygen concentration dropped very quickly in less than 10 days of storage and this can be partially explained by the emission of carbon monoxide. Depletion of oxygen is partly related to CO formation but a greater amount could be due to the radical-induced oxidative degradation of natural lipids, particularly the polyunsaturated linoleic acid [1]. A high energy of desorption required to overcome the bond indicates chemical adsorption of oxygen to the pellets. The auto-oxidation of fatty acids starts with formation of free radicals. In the presence of oxygen, hydroxyperoxide radicals are formed and in interaction with an unsaturated fatty acid produce two hydroxyperoxides and a new free radical. When pellets stored in air, hydroxyperoxides are formed from oxidation of fatty acids. Depending on whether oxygen bond or carbon bond break in them alcoxi radical, aldehydes, acids, hydrocarbons or ketones will be formed.

In order to examine the role of oxygen availability in the storage space on the rate of off-gassing, a set of experiments was conducted under controlled environment conditions. The same reactors described in small scale tests section were used in these experiments. Pellets were placed in 6 sealed reactors (2L by volume) that were purged with different gases. In two of the reactors pellets were stored in oxygen-free environments. When pellets were stored in an environment dominated by N_2 after purging, CO emission was as high as when pellets were stored under regular conditions. However when the reactors were purged with He (helium), the peak CO emission decreased to 25% of the values when pellets were stored under regular conditions. A hypothesis is that, although oxygen availability can accelerate the emission of

http://dx.doi.org/10.5772/58301

non–condensable gases, the O_2 content of the pellet material could be high enough to induce high amount of CO emission compared to pellets stored in air. Emission of carbon monoxide for pellets stored in CO_2 showed 50% less emission compared to pellets stored in air. Results obtained from stored pellets exposed to different head-space (HD) percentage showed that an increase in head-space is related to the peak emission factor for carbon monoxide and carbon dioxide with the following linear relationships (average values of two replicates each):

$$f_{CO} = 3 \times 10^{-4}[HD\%] + 0.0065$$

$$f_{CO_2} = 1.9 \times 10^{-3}[HD\%] + 0.0041$$

Oxygen concentration was also measured in all 13 locations including the head-space. Ever since the first gas sample was taken on day 2 of storage, the O_2 concentration decreased to 15% in the reactor. As seen in Figure 19, O_2 concentration was depleted rapidly to less than 5% within the first 7 days thus generating an oxygen-deficient atmosphere in a confined space. More oxygen was available within the bed of pellets compared to the head-space as a result of oxygen being trapped within the pellets until 40 days of storage when oxygen was consumed everywhere in the reactor. When oxygen levels fall below 19.5% by volume, air cannot support metabolism for an unlimited period of time. At 17% oxygen, the symptoms might simply be worse as reflected by hyper-ventilation. The oxygen-deficient atmosphere becomes more dangerous when oxygen content is further lowered to 15%; people can quickly progress to dizziness and rapid heartbeat. Finally, oxygen levels below 13% can lead to unconsciousness and eventually to death at around 6% oxygen [25].

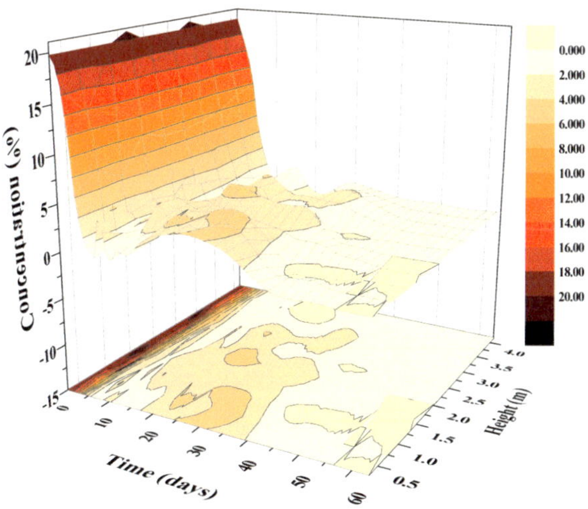

Figure 19. map of O_2 concentration in the silo at all locations (G0 to G13) during 63 days of storage [Concentrations are volume percentage]

5.4. 3D Analysis of radial distribution of emitted gases

Figure 20 shows CO_2 concentrations in the head-space and the bottom of the silo respectively. The gas concentration was measured at the same elevation but at 3 different radial positions (0.0, 0.6 and 1.2 m from the center) in order to investigate any possible radial concentration gradient. The results showed no significant variations in the CO_2 concentrations in different radial positions, due to the uniform radial temperature profile. The same measurements were made for all three gases (CO_2, CO and CH_4); again no significant differences in any of the gas concentrations were observed at different radial positions. However, Figure 20 clearly shows the difference in CO_2 concentrations along the axial locations, with the head-space concentration higher than that at the silo bottom which could be due to lower rate of mixing in the silo compared to the rate of reaction. The same observations were obtained for radial concentrations of carbon monoxide and methane over time.

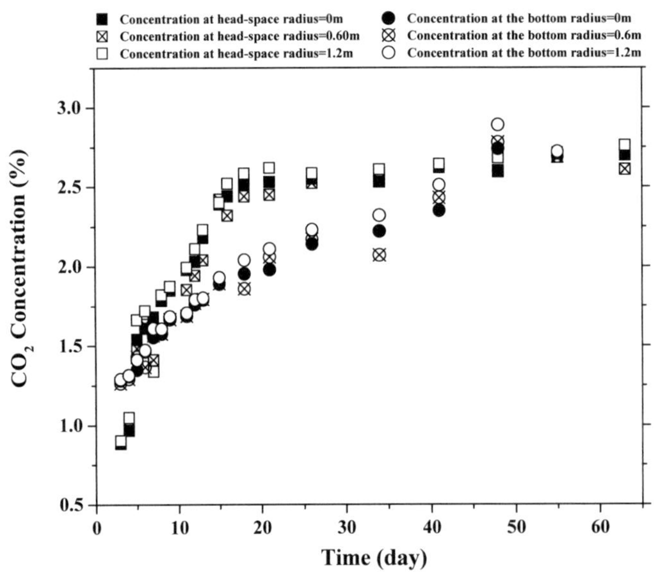

Figure 20. CO_2 concentration in the silo head-space and the bottom of the silo at different radial positions

6. Concluding remarks

The chapter has discussed various measurements of gases and volatiles, resulting data on emission concentrations and emission factors of gases in storages of wood pellets. The potential for flammability of gases emitted from wood pellets samples with moisture contents ranging from 4 to 50% stored in environment of 25 °C to 60 °C was investigated. The gas emission for CO_2, CO, CH_4 followed exponential profiles and the gas concentration was proportional to storage temperature, as expected. For samples with less than 15% MC at room temperature, CO_2 concentration increased up to ~1.7% which was the highest peak emission compared to even pellets with 35% and 50% MC. At higher temperatures, the maximum CO_2 concentration increased from 2.8% (40 °C, 4% MC) to 5.6% (40 °C, 50% MC) and from 9.6% (60 °C, 4% MC) to 10.9% (60 °C, 50% MC). For pellets with 4 and 15% MC, the concentration of CO increased as the temperature increased from 25 °C to 60 °C. However at a constant temperature of 25 °C, the peak concentration of CO was the same for wood pellets with 4% and 9% moisture contents and lower for pellets with 15%, 35% and 50% MC. Oxygen depletion was observed among all experimental conditions but showed to decrease more rapid for pellets with 9% MC (25 °C and 40 °C) compared to wood pellets with 4%, 15%, 35% and 50% MC. Least depletion of oxygen was observed for samples at 25 °C and 50% MC and accelerated as the temperature increased. In terms of the gas mixture ($CO/CH_4/H_2$), at higher temperatures (40 °C and 60 °C), the gas concentration decreased as the moisture content increased. Using ISO method for estimating the potential flammability of the gases emitted from the wood pellets, it can be concluded that gas concentrations does not reach flammable concentrations within the 5-week or even the 9-week testing period.

Emission and stratification of off-gasses from storage of wood pellets was also studied in pilot scale. The focus of the study was to investigate the spatial and temporal concentration of off-gases and purging effectiveness. Emission and stratification of off-gases was studied in pilot scale storage for over one year. Non-condensable gases that emitted from storage of material were carbon monoxide, carbon dioxide, methane and hydrogen. To study the stratification of gases, analysis of gas composition was done over time for different axial and radial positions. The emitted gases showed to have higher emission factor compared to work done with white wood pellets in small scale. It could be explained by the fact that off-gassing is a surface phenomenon and thus much active surface is available for reaction when larger amount of wood pellets are stored. Concentration of gases at plateau after 9 weeks of storage was 2.7 % CO_2, 0.14% CH_4 and 1.7% CO in the rector head-space. After a few days of storage some stratification were observed for carbon monoxide and methane. The clear stratification of carbon monoxide could be due to high uptake of CO by wood pellets over time. The unstable environment in the storage of wood pellets in the first few weeks of storage was due to very rapid consumption of oxygen in the space, reaction of emitted CO and CO_2 with pellets, development of high temperature spots within the bulk due to higher activity of pellets and thus higher local temperatures, and moisture migration within the bed. Oxygen plays an important role in accelerating the emission of carbon monoxide. Results from experiment where material was kept in oxygen-free environment (N_2-rich) confirmed the same lethal concentration of CO possibly through onsuming the oxygen within the material.

Apendix A

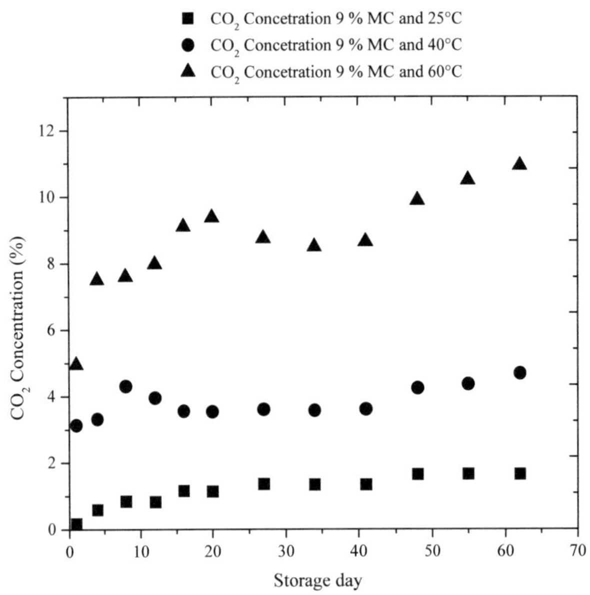

Figure A 1. Concentration of CO_2 for wood pellet with 9% moisture content over 62 days at 25°C, 40°C and 60°C

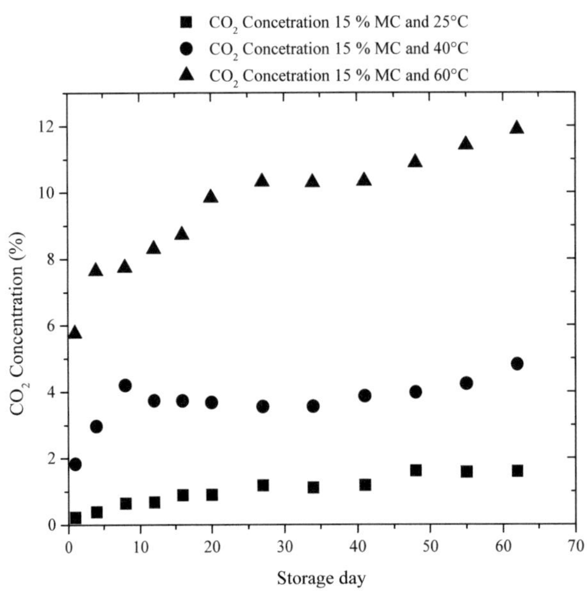

Figure A 2. Concentration of CO_2 for wood pellet with 15% moisture content over 62 days at 25°C, 40°C and 60°C

http://dx.doi.org/10.5772/58301

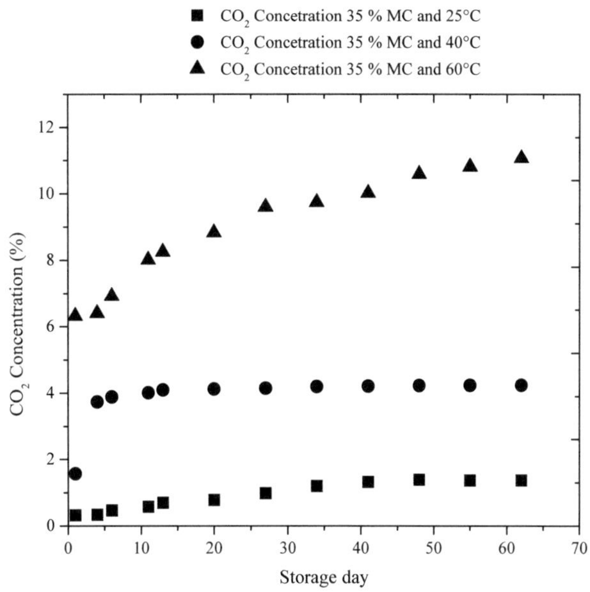

Figure A 3. Concentration of CO_2 for wood pellet with 35% moisture content over 62 days at 25°C, 40°C and 60°C

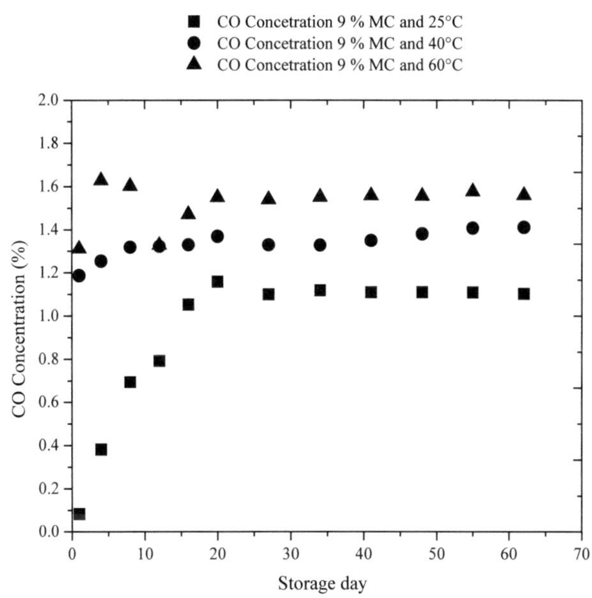

Figure A 4. Concentration of CO for wood pellet with 9% moisture concentration over 62 days at 25°C, 40°C and 60°C

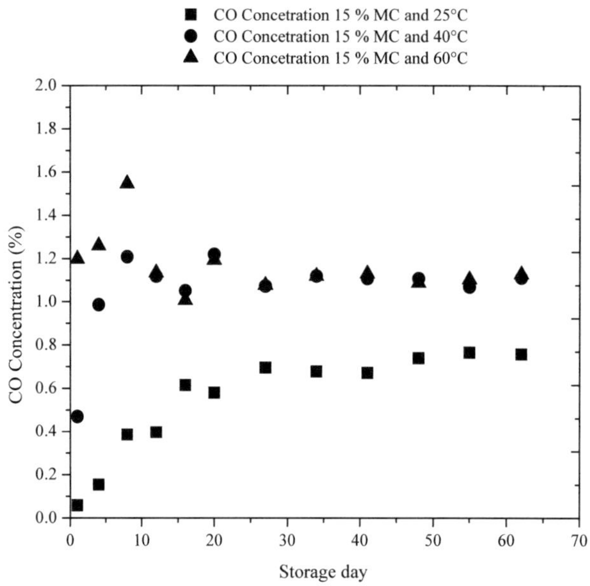

Figure A 5. Concentration of CO for wood pellet with 15% moisture concentration over 62 days at 25°C, 40°C and 60°C

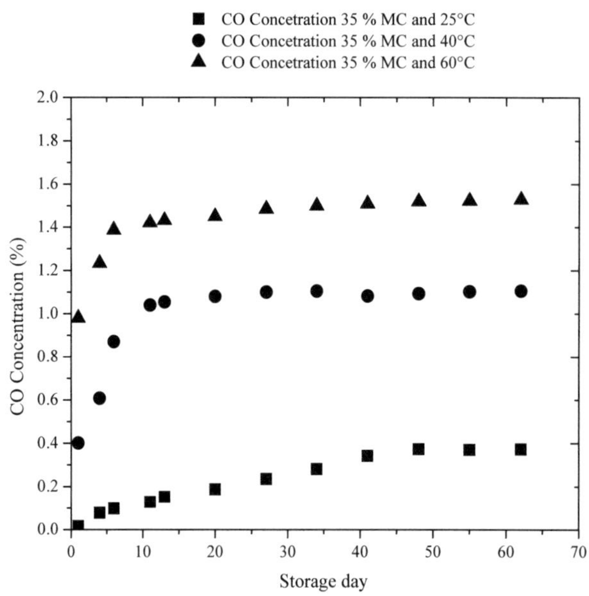

Figure A 6. Concentration of CO for wood pellet with 35% moisture content over 62 days at 25°C, 40°C and 60°C

http://dx.doi.org/10.5772/58301

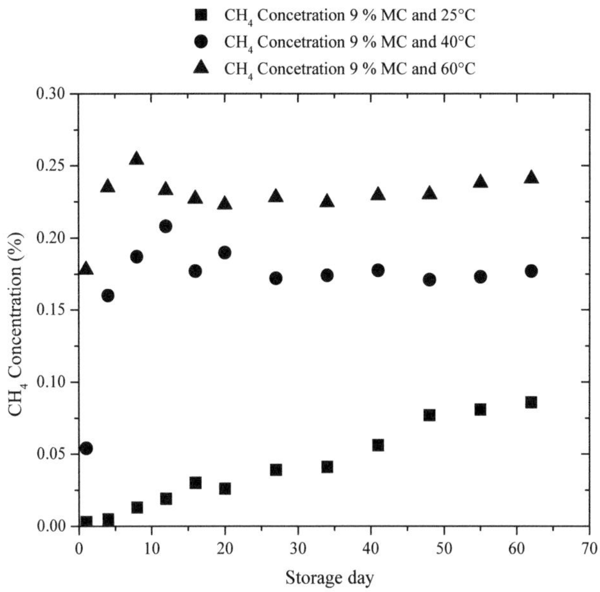

Figure A 7. Concentration of CH_4 for emissions from wood pellet with 9% moisture content over 62 days at 25°C, 40°C and 60°C

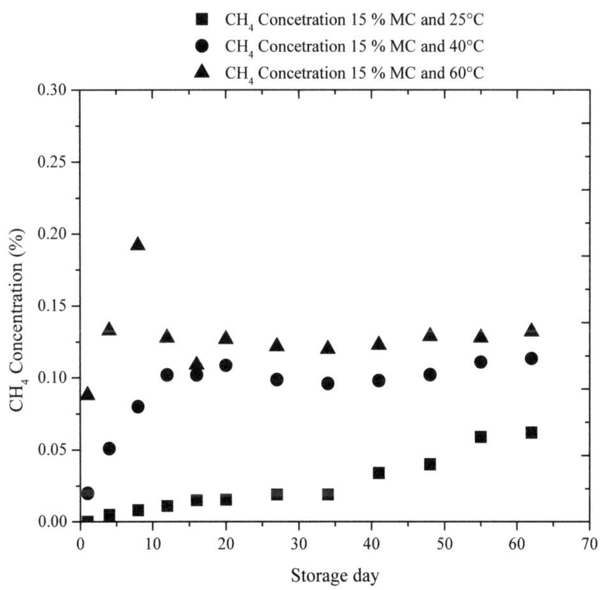

Figure A 8. Concentration of CH_4 for emissions from wood pellet with 15% moisture content over 62 days at 25°C, 40°C and 60°C

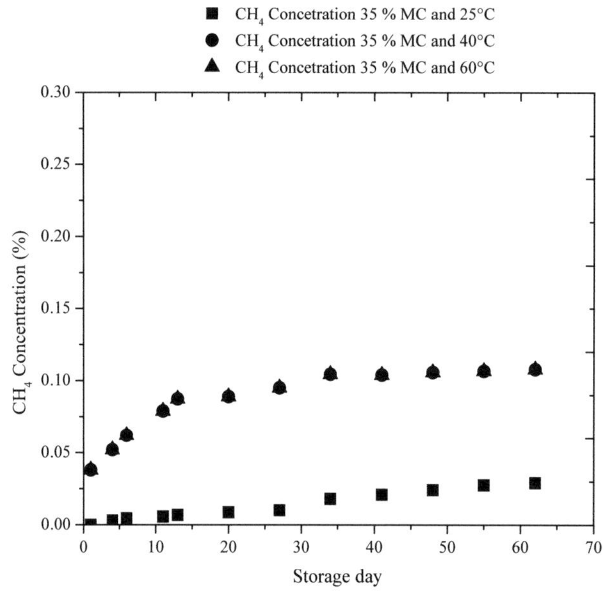

Figure A 9. Concentration of CH_4 for emissions from wood pellet with 35% moisture content over 62 days at 25°C, 40°C and 60°C

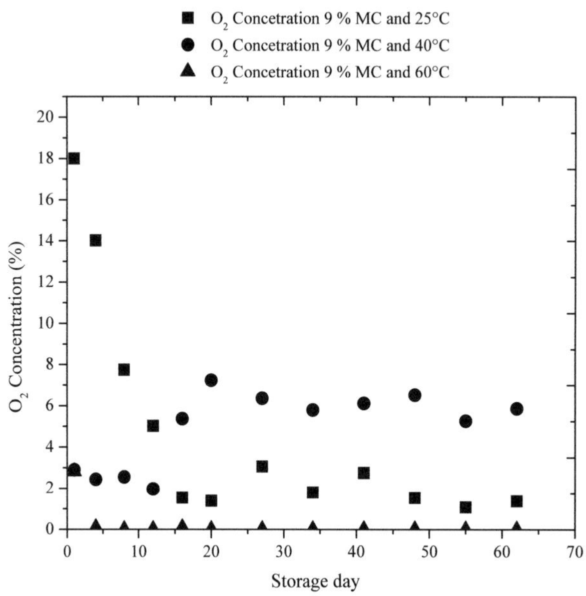

Figure A 10. Oxygen depletion for wood pellet with 9% moisture content over 62 days at 25°C, 40°C and 60°C

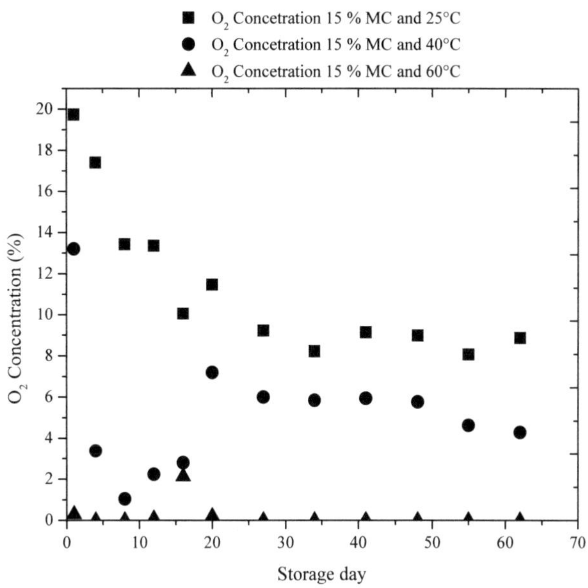

Figure A 11. Oxygen depletion for wood pellet with 15% moisture content over 62 days at 25°C, 40°C and 60°C

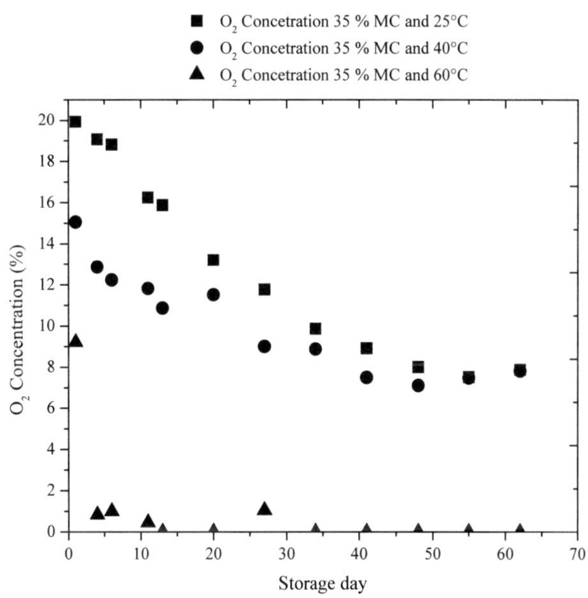

Figure A 12. Oxygen depletion for wood pellet with 35% moisture content over 62 days at 25°C, 40°C and 60°C

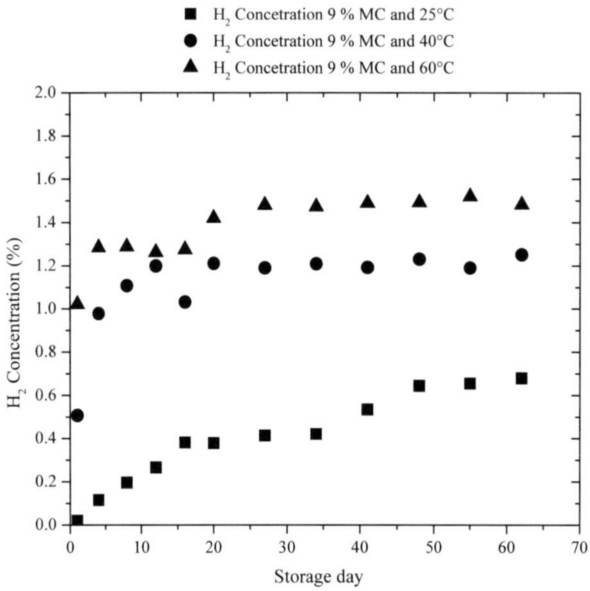

Figure A 13. Concentration of hydrogen for emissions from wood pellet with 9% moisture content over 62 days at 25°C, 40°C and 60°C

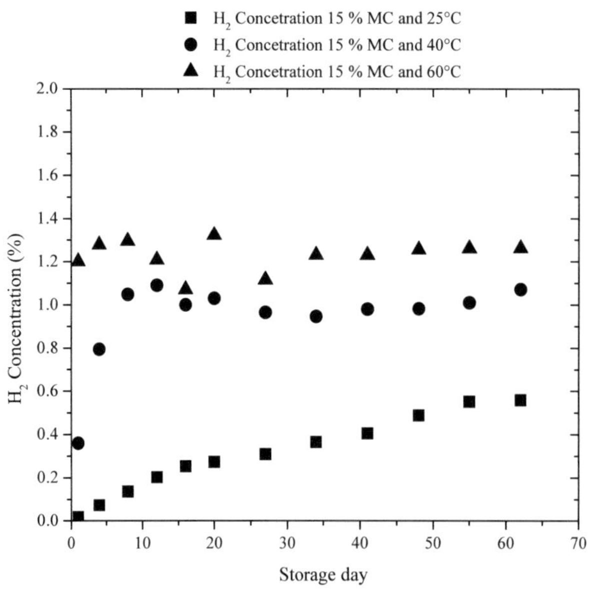

Figure A 14. Concentration of hydrogen for emissions from wood pellet with 15% moisture content over 62 days at 25°C, 40°C and 60°C

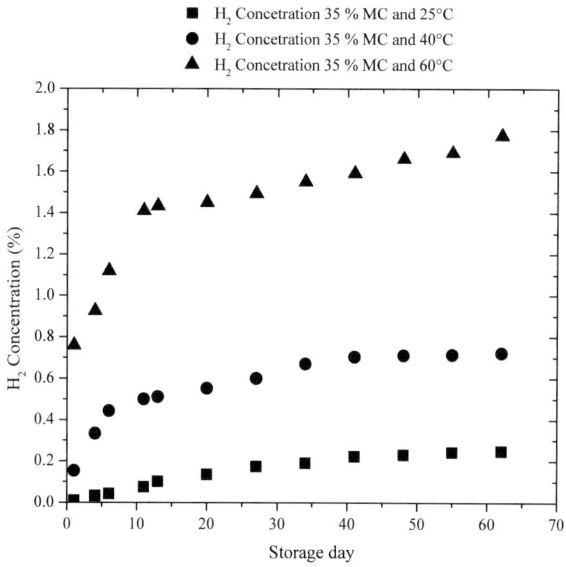

Figure A 15. Concentration of hydrogen for emissions from wood pellet with 35% moisture content over 62 days at 25°C, 40°C and 60°C

Acknowledgements

This research was funded in parts by the Wood pellet Association of Canada, Natural Sciences and Engineering Research Council of Canada and Oak Ridge National Laboratory. Thanks are also extended to Premium Pellet Ltd. and Fibreco Export Inc. and Pinnacle Renewable Energy Group, Inc. for donating wood pellets for this study.

Author details

Fahimeh Yazdanpanah[1*], Shahab Sokhansanj[1], Hamid Rezaei[1], C. Jim Lim[1], Anthony K. Lau[1], X. Tony Bi[1], S. Melin[2], Jaya Shankar Tumuluru[3] and Chang Soo Kim[4]

*Address all correspondence to: yazdanpanah@gmail.com

1 Chemical & Biological Engineering, University of British Columbia, Vancouver, Canada

2 Wood pellet Association of Canada, Vancouver, Canada

3 Bioenergy Group, Idaho National Laboratory, Idaho Falls, USA

4 Korea Advanced Institute of Science and Technology, Daejeon, South Korea

References

[1] Kuang X, Shankar TJ, Bi XT, Sokhansanj S, Lim CJ, Melin S. Characterization and kinetics study of off-Gas emissions from stored wood pellets. Ann Occup Hyg 2008; 52: 675-683.

[2] Tumuluru J, Kuang X, Sokhansanj S, Lim CJ, Bi T, Melin S. Development of laboratory studies on the off-gassing of wood pellets. Canadian Biosystems Engineering 2010; 52: 8.1,8.9.

[3] Guo W. Self heating and spontaneous combustion of wood pellets during storage. PhD Dissertation, University of British Columbia 2013.

[4] Hagström K. Occupational exposure during production of wood pellets in Sweden. Doctoral Dissertation 2008.

[5] Svedberg URA, Hogberg HE, Hogberg J, Galle B. Emission of hexanal and carbon monoxide from storage of wood pellets, a potential occupational and domestic health hazard. Ann Occup Hyg 2004; 48: 339-349.

[6] Arshadi M, Gref R. Emission of volatile organic compounds from softwood pellets during storage. For Prod J 2005; 55: 132-135.

[7] Risholm-Sundman M, Lundgren M, Vestin E, Herder P. Emissions of acetic acid and other volatile organic compounds from different species of solid wood. Holz Als Roh-Und Werkstoff 1998; 56: 125-129.

[8] Wihersaari M. Evaluation of greenhouse gas emission risks from storage of wood residue. Biomass & Bioenergy 2005; 28(5): 444-453.

[9] Wihersaari M. Greenhouse gas emissions from final harvest fuel chip production in Finland. Biomass & Bioenergy 2005; 28: 435-43.

[10] Manninen A, Pasanen P, Holopainen JK. Comparing the VOC emissions between air-dried and heat-treated Scots pine wood. Atmos Environ 2002; 36: 1763-1768.

[11] Banerjee S. Mechanisms of terpene release during sawdust and flake drying 2001; 55: 413-416.

[12] Roffael E. Volatile organic compounds and formaldehyde in nature, wood and wood based panels. Holz Als Roh-und Werkst 2006; 64: 144-149.

[13] Rupar K, Sanati M. The release of terpenes during storage of biomass. Biomass Bioenerg 2005; 28: 29-34.

[14] McGraw GW, Hemingway RW, Ingram LL, Canady CS, McGraw WB. Thermal degradation of terpene: camphene, delta-3-carenic, limonene and alfa-terpinene. Environmental Science & Technology 1999; 33: 4029-4033.

[15] Granstrom KM. Emissions of Hexanal and Terpenes during Storage of Solid Wood Fuels. For Prod J 2010; 60: 27-32.

[16] Melin S, Svedberg U, Samuelsson J. Emission from wood pellets during ocean transportation 2008; WPAC Research Report.

[17] Kuang X, Shankar TJ, Bi XT, Lim CJ, Sokhansanj S, Melin S. Rate and peak concentrations of off-gas emissions in stored wood pellets-sensitivities to temperature, relative humidity and headspace volume. Ann Occup Hyg 2009; 53: 789-796.

[18] Kuang X, Shankar TJ, Sokhansanj S, Lim CJ, Bi XT, Melin S. Effects of headspace and oxygen level on off-gas emissions from wood pellets in storage. Ann Occup Hyg 2009; 53: 807-813.

[19] Yazdanpanah F. Evolution and stratification of off-gasses in stored wood pellets. PhD Dissertation, University of British Columbia 2013.

[20] Tumuluru JS, Sokhansanj S, Lim CJ, Bi X, Meli S, Swaan J. Design and Instrumentation of a Large Column Container to Study the Off-Gassing, Gas Stratification and Physical Properties of Wood Pellets during Storage. ASABE Annual General Meeting 2009.

[21] ISO 10156 Standard. Gases and gas mixtures – Determination of fire and oxidizing ability for selection of cylinder valve outlets 2010.

[22] US. Department of Labor. Occupational Safety and Health Administration (www.osha.gov).

[23] Carpenter TM, Fox EL. Absence of stratification and rapidity of mixing of carbon dioxide in air samples. J Biol Chem 1927; 73:379-381.

[24] Theilacker JC, White MJ. Diffusion of gases in air and its affect on oxygen deficiency hazard abatement. Transactions of the cryogenic Engineering conference - CEC 2006; 51.

[25] Koppejan J, Lönnermark A, Persson H, Larsson I, Blomqvist P, Arshadi M, et al. Health and safety aspects of solid biomass storage, transportation and feeding. IEA Bioenergy 2013.